AF864139

TRANSPORTATION INFRASTRUCTURE - ROADS, HIGHWAYS, BRIDGES, AIRPORTS AND MASS TRANSIT

EMERGENCY RELIEF PROGRAMS FOR U.S. HIGHWAYS AND ROADS

ELEMENTS AND CONSIDERATIONS

TRANSPORTATION INFRASTRUCTURE - ROADS, HIGHWAYS, BRIDGES, AIRPORTS AND MASS TRANSIT

Additional books in this series can be found on Nova's website
under the Series tab.

Additional e-books in this series can be found on Nova's website
under the e-books tab.

TRANSPORTATION ISSUES, POLICIES AND R&D

Additional books in this series can be found on Nova's website
under the Series tab.

Additional e-books in this series can be found on Nova's website
under the e-books tab.

TRANSPORTATION INFRASTRUCTURE - ROADS, HIGHWAYS, BRIDGES, AIRPORTS AND MASS TRANSIT

EMERGENCY RELIEF PROGRAMS FOR U.S. HIGHWAYS AND ROADS

ELEMENTS AND CONSIDERATIONS

CODY O. HARRISON
AND
TED J. GIBSON
EDITORS

Nova Science Publishers, Inc.
New York

Copyright © 2012 by Nova Science Publishers, Inc.

All rights reserved. No part of this book may be reproduced, stored in a retrieval system or transmitted in any form or by any means: electronic, electrostatic, magnetic, tape, mechanical photocopying, recording or otherwise without the written permission of the Publisher.

For permission to use material from this book please contact us:
Telephone 631-231-7269; Fax 631-231-8175
Web Site: http://www.novapublishers.com

NOTICE TO THE READER

The Publisher has taken reasonable care in the preparation of this book, but makes no expressed or implied warranty of any kind and assumes no responsibility for any errors or omissions. No liability is assumed for incidental or consequential damages in connection with or arising out of information contained in this book. The Publisher shall not be liable for any special, consequential, or exemplary damages resulting, in whole or in part, from the readers' use of, or reliance upon, this material. Any parts of this book based on government reports are so indicated and copyright is claimed for those parts to the extent applicable to compilations of such works.

Independent verification should be sought for any data, advice or recommendations contained in this book. In addition, no responsibility is assumed by the publisher for any injury and/or damage to persons or property arising from any methods, products, instructions, ideas or otherwise contained in this publication.

This publication is designed to provide accurate and authoritative information with regard to the subject matter covered herein. It is sold with the clear understanding that the Publisher is not engaged in rendering legal or any other professional services. If legal or any other expert assistance is required, the services of a competent person should be sought. FROM A DECLARATION OF PARTICIPANTS JOINTLY ADOPTED BY A COMMITTEE OF THE AMERICAN BAR ASSOCIATION AND A COMMITTEE OF PUBLISHERS.

Additional color graphics may be available in the e-book version of this book.

LIBRARY OF CONGRESS CATALOGING-IN-PUBLICATION DATA

ISBN: 978-1-61942-858-4

Published by Nova Science Publishers, Inc. † New York

CONTENTS

PREFACE

In recent years, many states have experienced natural disasters, such as hurricanes, floods, and storms, which have caused catastrophic damage to transportation infrastructure and overwhelmed the capacity of state and local governments to respond and recover. Reconstruction after these events can cost taxpayers billions of dollars. As part of the continuing federal role in responding to and recovering from natural disasters and similar events, the Federal Highway Administration (FHWA), within the U.S. Department of Transportation (DOT), administers the Emergency Relief Program which provides funding to repair or reconstruct federal-aid highways and roads on federal lands damaged or destroyed by natural disasters and other catastrophic events. This book provides an overview of emergency relief programs for U.S. highways and roads with a focus on strengthening oversight of project eligibility decisions.

In: Emergency Relief Programs for U.S. Highways and Roads ISBN: 978-1-61942-858-4
Editors: Cody O. Harrison and Ted J. Gibson © 2012 Nova Science Publishers, Inc.

Chapter 1

EMERGENCY RELIEF MANUAL: (FEDERAL-AID HIGHWAYS)*

Office of Infrastructure

I. INTRODUCTION

A. Purpose of Manual

This manual is an update of the Emergency Relief Manual, Interim Update August 2003. It provides updated guidance and instructions on the Federal Highway Administration's (FHWA) emergency relief (ER) program. This manual provides information for FHWA, State, and local transportation agency personnel on policies and procedures for requesting, obtaining and administering ER funds. This manual covers only those criteria and procedures applicable to the ER program for Federal-aid highways, these being public highways other than those functionally classified as local roads or rural minor collectors. Procedures relating to the ER program for roads on Federal lands that are not Federal-aid highways are outlined in *Emergency Relief for Federally Owned Roads (ERFO) Disaster Assistance Manual, Publication Number FHWA-FLH-04-007*. Damage to highway facilities that are neither Federal-aid highways nor roads on Federal lands may be eligible for other Federal funds authorized by the Stafford Act, P. L. 93- 288, administered by the Federal Emergency Management Agency (FEMA). The FEMA publication, *Public Assistance Guide, June 2007*, provides a detailed reference of the Public Assistance Program requirements and procedures.

B. Program Overview

Congress authorized in Title 23, United States Code, Section 125, a special program from the Highway Trust Fund for the repair or reconstruction of Federal-aid highways and roads on

* This is an edited, reformatted and augmented version of an Office of Infrastructure, Office of Program Administration and Federal Highway Administration publication, dated November 2009.

Federal lands which have suffered serious damage as a result of (1) natural disasters or (2) catastrophic failures from an external cause. This program, commonly referred to as the emergency relief or ER program, supplements the commitment of resources by States, their political subdivisions, or other Federal agencies to help pay for unusually heavy expenses resulting from extraordinary conditions.

Examples of natural disasters include floods, hurricanes, earthquakes, tornadoes, tidal waves, severe storms, and landslides. A catastrophic failure is defined as the sudden and complete failure of a major element or segment of the highway system that causes a disastrous impact on transportation services. The failure must be catastrophic in nature. Additionally, in order to be eligible for ER, the cause of the failure must be determined to be external to the facility. Both conditions must be satisfied. A bridge suddenly collapsing after being struck by a barge is an example of a catastrophic failure from an external cause.

ER funds are not intended to cover all damage repair costs nor interim emergency repair costs that will necessarily restore the facility to pre-disaster conditions. State and local highway agencies must expect additional expenditures, changes in project priorities, and some inconvenience to traffic as a result of emergency conditions. State and local governments are responsible for planning and providing for extraordinary conditions. Economic hardship is not a factor in determining repair eligibility. Although there is no nationwide definitive monetary break point between what is considered routine and extraordinary repair expenses, the FHWA has determined that eligible ER repair activities in a State in the range of $700,000 (Federal share) or more are usually significant enough to justify approval of ER funds. The $700,000 threshold applies to Federal-aid highway damage eligible under the ER program separate from Federally-owned highway damages eligible under the ERFO program. FHWA's Office of Federal Lands Highway imposes a separate $700,000 threshold for eligible ERFO events on Federally-owned lands.

State and local governments are expected to prepare for certain natural events such as rainfall and flooding through adequate planning and commitment of resources. Additionally, known recurring or seasonal weather patterns or conditions should be planned for through a State or local agency's annual maintenance budget. Damages from the occurrence of such natural events, other than those of an extraordinary nature, should not rely on the ER program for assistance.

By law, the FHWA can provide up to $100 million in ER funding for each natural disaster or catastrophic failure event within a State that is found eligible for funding under the ER program (commonly referred to as the $100 million per State cap). Because of the limited amount of money authorized annually for the ER program and the likelihood that a number of states will experience ER events, funding for large events is likely to be provided over a two (or more) year time period. Also, the total ER obligation for US Territories (American Samoa, Commonwealth of Northern Mariana Islands, Guam, and Virgin Islands) is limited to $20 million in any fiscal year (FY). For a large disaster that exceeds the $100 million per State cap, Congress may pass special legislation lifting the cap for that disaster.

Note that the $100 million per State cap applies to the combined cost of damages to both Federal-aid highways (ER) and Federal Lands highways (ERFO) for any single event within a State.

ER funds are available for permanent repairs and for work accomplished more than 180 days after an event at the pro rata Federal-aid share that would normally apply to the Federal-aid facility being repaired. For Interstate highways, the Federal share is 90 percent. For all other highways, the Federal share is 80 percent. The Federal share can increase in States with high percentages of Federally owned public lands (known as "sliding scale rates"). *Emergency repair work to restore essential traffic, minimize the extent of damage, or protect the remaining facilities, accomplished in the first 180 days after the occurrence of the disaster, may be reimbursed at 100 percent Federal share.* During this 180-day period, permanent repair work is reimbursed at the normal pro rata share even if the permanent repair is performed as an incidental part of the emergency repair work. Permanent repair work is not to be considered emergency repair work for the purpose of establishing the eligible Federal share, and can only reimbursed at 100% if special legislation allows.

The applicability of the ER program to a natural disaster is based on the extent and intensity of the disaster. Damage to highways must be severe, occur over a wide area, and result in unusually high expenses to the highway agency. The ER program also applies to catastrophic failures (sudden and complete failures due to an external cause) and which result in a disastrous impact on transportation services and unusually high expenses to the highway agency. Failures due to an inherent flaw in the facility itself do not qualify for ER assistance.

For natural disasters, Federal interagency coordination is handled through an interagency agreement between the FEMA and 11 other Federal agencies with hazard mitigation responsibilities. Hazard mitigation teams are activated immediately following a disaster. The FEMA's publication *"Flood Hazard Mitigation Handbook of Common Practices"* documents the appropriate activity.

In general, the FHWA Division Administrator must evaluate a Damage Survey Summary Report, discussed in Chapter V, and make a finding that a disaster is eligible within the intent of the law and applicable regulations before ER funds can be made available. A limited amount of funding may be made available under the "Quick Release" method outlined in Chapter III. The role and responsibilities of the FHWA in ER activities under 23 USC. Sections 120 and 125 are:

1) Administration of the ER program through coordination and implementation of disaster relief policies and procedures.
2) Assistance to State, Federal, or other highway agencies in applying for funds.
3) Technical assistance to the State, Federal, or other highway agencies in the review, design, repair, and reconstruction of damaged highway facilities.

II. Eligibility of Damage Repair Work

A. General

Roadways and bridges that are on a Federal-aid highway and that are damaged as a direct result of an approved natural disaster or catastrophic failure from an external cause are eligible for ER funds. A Federal-aid highway is defined in title 23 United States Code (23 USC.) section 101 as a "highway eligible for assistance under this chapter other than a highway classified as a local road or rural minor collector." Therefore, any highway that is not eligible for assistance under the provisions of 23 USC. Chapter 1, are not eligible for Federal-aid highway ER funding. As a general rule, items that are eligible for participation as a part of a regular Federal-aid improvement project under title 23 may be eligible under the ER program.

Normally, eligible work must be within the right-of-way limits of the damaged Federal-aid highway facility. A minimum $5,000 in repair cost per site (refer to Chapter II, Section C-2, *Damage Estimate Under $5,000 per Site*) should be used to determine if the extent of repair work at a site is beyond the scope of heavy maintenance. Title 23 Code of Federal Regulations (23 CFR) Part 668, Subpart A includes a $700,000 (Federal share) disaster eligibility threshold to distinguish between heavy maintenance or routine emergency repair and serious damage eligible under the ER program.

The ER program provides for repair and restoration of highway facilities to pre-disaster conditions. Restoration *in kind* is therefore the predominate type of repair expected to be accomplished with ER funds. ER funds are not intended to replace other Federal-aid, State, or local funds for new construction to increase capacity, correct non-disaster related deficiencies, or otherwise improve highway facilities.

Added protective features, such as the relocation or rebuilding of roadways at higher elevation or lengthening or raising bridges, and added facilities not existing prior to the natural disaster or catastrophic failure, such as additional lanes, upgraded surfacing, or structures are commonly referred to as a betterment. Betterments are not generally eligible for ER funding unless justified. The eligibility of betterments is discussed in more detail later in this chapter.

All repair work falls under two major categories: 1) emergency repairs and 2) permanent repairs. *Emergency repairs* are those repairs during and immediately following a disaster to restore essential traffic, to minimize the extent of damage, or to protect the remaining facilities. These repairs can begin immediately following a disaster, and prior FHWA approval is not required. Properly documented costs will later be reimbursed once the FHWA Division Administrator makes a finding that the disaster is eligible for ER funding. *Permanent repairs* are those repairs undertaken (usually after emergency repairs have been completed) to restore the highway to its pre-disaster condition. Permanent repairs must have prior FHWA approval and authorization unless done as part of the emergency repairs.

It is important to understand the difference in the Federal share for these two categories of repairs. For emergency repairs, the Federal share is 100 percent for repair work done to restore essential traffic, to minimize the extent of damage or to protect the remaining facilities within the first 180 days after the occurrence of the disaster. On the other hand, for permanent repairs, the Federal share depends on the type of Federal-aid highway being repaired. For Interstate highways, the Federal share is 90 percent. For all other Federal-aid highways, the Federal share is 80 percent. The Federal share may be increased in states with a high percentage of Federally owned public lands. A more detailed discussion of emergency versus permanent repair follows later in this chapter. Regardless of whether the work is considered emergency repairs or permanent repairs, the Federal share is 100 percent for work done on roads on Federal lands.

B. Eligible Items

Generally, all elements of the highway within its cross section damaged as a direct result of a disaster are eligible for repair under the ER program. This includes, but is not limited to, elements such as pavement, shoulders, slopes and embankments, guardrail, signs and traffic control devices, bridges, culverts, cribbing or other bank control features, bike and pedestrian path, fencing, and retaining walls. When a pedestrian or bicycle trail that is within the right-of-way of a Federal-aid highway is damaged, that damage is eligible for ER funding whether or not the roadway itself is damaged.

The intent of the ER program is to fund repairs to damaged roadways caused by a natural disaster or catastrophic failure, not repairs to roadways damaged as a result of preexisting and non-disaster related, i.e., inherent deficient conditions. The following is a more detailed discussion of eligibility for various highway elements:

1. Engineering and Right-of-Way

Preliminary engineering, right-of-way, and construction engineering directly attributable to repair of eligible damage are eligible for ER reimbursement. Reasonable construction engineering costs are eligible. Maintenance, administration, and overhead costs of State or local governments and of other Federal agencies are not eligible. Costs such as a general overall assessment of damage, general supervision, contract administration other than construction engineering, and project planning and scheduling are considered indirect costs that are eligible for ER funding (see *Indirect Costs* below).

2. Indirect Costs

Section 1212(a) of the Transportation Equity Act for the 21st Century (TEA-21), amended section 302 of 23 USC. to allow State transportation departments to claim reimbursement of indirect costs. State and local government agencies can be reimbursed for indirect costs subject to the provisions of the Office of Management and Budget Circular A-87 (OMB Circular), "Cost Principles for State, Local, and Indian Tribal Governments." Since grantees are not required to claim indirect costs, these procedures only apply to agencies otherwise seeking reimbursement of these costs.

Federal-aid highway program funds may participate in the indirect costs of State and local governments when the costs are properly allocated to all benefiting cost objectives in accordance with the provisions of the OMB Circular.

By regulation, 2 CFR 225 Appendix E points out the need to properly account for "extraordinary or distorting expenditures" in order to ensure an equitable distribution of indirect costs to all benefiting cost objectives (Federal and non-Federal awards/activities). Not making allowances for the one-time infusion of significant amounts of Federal dollars into the Federal-aid Highway Program (such as ER funding following a disaster) will likely result in a significant over-recovery of indirect costs that will have to be recovered at a later time.

The State Transportation Agency should develop an indirect cost rate that equitably allocates indirect costs to ER projects. A separate indirect cost pool and direct program pool or pools, should be established for calculation of an indirect cost rate associated with administering ER activities. A revised Indirect Cost Allocation Plan, should be submitted to the FHWA Division Office for review and approval.

For additional guidance concerning indirect costs, please refer to FHWA Memorandum "Clarification of Policy on Indirect Costs of State and Local Governments," May 5, 2004 (available online at http://www.fhwa.dot. gov/ legsregs/directives/policy/indirectcost.htm).

3. Detours

Subject to FHWA review and approval, designated detours and supplemental detours are eligible if such detour routes can be shown to relieve excess traffic directly attributable to the eligible disaster. Such eligible items may include additional traffic control, necessary overlays, and required widening needed to support the excess traffic.

Temporary connections should be constructed consistent with anticipated volume and type of essential traffic, reasonable speeds, safety, term of use, and costs. However, ER funding does not necessarily extend to restoring pre-disaster levels of service until completion of permanent repairs.

Where a temporary structure or an alternate existing route is not reasonable or practical as a temporary connection, construction of a temporary ferry service (ferryboat, ferry operation and maintenance, docking and loading facilities) is eligible for ER funds. ER participation in the ferryboat is limited to acquisition costs, less resale value, or to a reasonable rental fee. Although passenger-only ferry services are eligible for ER participation, bus transit services to transport passengers to and from the ferry terminal would not be an eligible ER expense.

Repair of surface damage to a designated detour caused by traffic that has been detoured from a damaged or impassable Federal-aid highway is eligible for ER funds. This may include roadway surface repairs to provide reasonable traffic service during the period of time the detour is in use as well as surface repairs to the detour route to restore the detour route to pre-disaster condition after detour traffic has been removed. A documented survey of the condition of the proposed designated detour route prior to detouring traffic on it should be helpful in determining the scope of restoration work needed to restore the roadway to its pre-disaster condition. A designated detour, which also could be a non-Federal-aid highway, is defined as the officially signed detour that highway officials have established to reroute traffic around the damaged or impassable portion of a Federal-aid highway.

4. Traffic Damage

In general, repair of roadway surfaces due to traffic damages, even if damage is aggravated by saturated subgrade conditions, or by inundation of the roadway, is not eligible for ER funds. There are three exceptions. ER funds may participate in repair of surface damage to:

1) Any public roads - caused by vehicles making repairs to Federal-aid highways,
2) Any public roads - caused by traffic using the officially designated detour around a damaged Federal-aid highway, and
3) Any Federal-aid highways - caused by vehicles responding to a disaster.

ER participation should be limited to surface damage that has occurred during the first 60 days after a disaster event, unless otherwise approved by the FHWA Division Administrator. Examples of response activities include vehicles involved in repairing other transportation facilities, constructing emergency dikes or performing emergency repairs to dikes, providing essential services such as fire fighting and providing supplies, or removing debris from both public and private property. In exceptional cases, ER participation may be extended to damage occurring up to 6 months after a disaster.

Identifying surface damage caused by response vehicles can be difficult. In some cases, to aid in the decision, an operational pavement management system might provide data that would confirm the roadway condition prior to the disaster. In most cases, an analysis based on best professional engineering practice will be needed. This analysis should use other appropriate information on the pre-disaster condition of the roadway surface and the special circumstances and vehicle use that caused the damage. Consequently, FHWA Division Office personnel should field-visit all sites involved in this eligibility category and provide a written report of observed conditions.

5. Overlays

Where entire sections of roadways are damaged and need to be reconstructed, new surfacing is eligible. In addition, where several intermittent but close-by sites need to have the surfacing repaired, resurfacing of the entire section between and including the sites is eligible.

Overlays of roadways that, even though submerged during the flooding, have suffered no significant damage as a direct result of the flooding, are not eligible.

6. Raising Grades

a. Traditional Flooding

Temporary work to raise roadway grades to maintain essential traffic service during the flooding is eligible. This is limited to fill material and minimum riprap to protect the temporary fill plus temporary surfacing material. If such roadways have otherwise suffered no significant damage as a direct result of the flood, work to provide a permanent higher grade, e.g., re-compact fill, provide permanent surfacing, provide drainage, guardrail, signing, etc., is not eligible.

Where roadways have been severely damaged and substantial lengths need to be permanently reconstructed, raising the roadway grade as a part of the permanent reconstruction project to avoid future flooding problems could be eligible, on a case-by-case basis, if determined to be a cost-effective betterment. Again, raising the grade of roadways that, even though submerged during the flooding, have suffered no significant damage as a direct result of the flooding, is not eligible.

b. Basin Flooding

A basin is a large depression in the land with no natural drainage outlet such that the water level decreases only due to infiltration or evaporation.

ER funding is available to raise the grades of *critical* Federal-aid highways faced with *long-term* loss of use due to an *unprecedented* rise in basin water level when basin flooding is considered a natural disaster for the purpose of the ER program, and if the corrective work is restorative rather than preventive in nature.

Basin flooding is considered an eligible disaster under the ER program if it can be shown that:

i. The work is restorative rather than preventive. A restorative situation exists if damage was not anticipated, and the flooding is long-term, effectively rendering the roadway useless for an extended period of time.

ii. There has been an *unprecedented* rise in the basin water level, both in terms of the magnitude of the increase and the time frame in which this increase occurred. To document the unprecedented rise, the State should provide information showing that the water elevation in the basin has reached historically high levels. Further, it should provide information that the rise in the water level occurred during a short period of time such that the rate of increase was much greater than previously experienced. Support information could include, for example, historic water level elevation and rainfall intensity records. If these are not available for the basin, more empirical evidence, such as State and/or local maintenance reports or other information that provides some historical perspective on events and water levels within the basin, could be used.

iii. There is severe damage that results in *long-term* loss of use of *critical* Federal-aid routes.

iv. To document the long-term loss, information should be provided concerning the length of time highway facilities have been closed to traffic and are projected to remain closed to traffic based on the basin water level elevations that have and/or are projected to occur.

In the case of major arterials still in service which the State determines must remain open, documentation on long-term loss of use will have to be based on the potential for this loss to occur should anticipated basin water levels be reached. Since the ER program is not of a preventative nature but only provides funding after a disaster has occurred, extreme care is needed in evaluating situations involving potential loss of use to ensure the integrity of the ER program is maintained.

Only those routes that are critical to restoring traffic service are eligible for grade raises. Factors to take into account in evaluating whether individual routes are critical could include functional classification, provision of essential community services such as access for school, ambulance, fire, and mail vehicles, availability of alternate routes, length of detours, etc. The FHWA Division Office in cooperation with the State, will jointly determine those critical Federal-aid routes eligible for grade raises.

Basin flooding is handled as a separate disaster under the ER program. Generally, each basin will be evaluated individually based on the water level rise that has occurred within that basin, and the above criteria must be satisfied for that basin before ER funding will be considered. However, special cases may arise where several basins in close proximity, all of which are affecting the same critical Federal-aid routes, may be treated as one event for the purposes of an ER finding that a disaster has occurred.

The beginning date for incurring ER eligible costs for a basin flooding disaster is not as clear-cut as the typical ER disaster. Basin flooding may represent the culmination of several meteorological events that have caused excessive run-off into the basin and have occurred over an extended period of time. For a typical ER eligible disaster, a State's letter of intent for a basin flooding disaster will serve as the beginning date of ER eligibility for that disaster. If a Division Office believes that use of another beginning date requested by the State can be justified, the damage survey summary report accompanying a State's request for ER funding for a basin flooding disaster must present their recommendation and rationale as to an alternate beginning date, with appropriate available supporting information. In an effort to provide some consistency on this matter, Headquarters must be consulted for use of an alternate date other than the letter of intent date for a basin flooding disaster.

The Federal share for ER funding provided for basin flooding should be determined in a manner similar to other ER disasters. For a basin flooding disaster, costs to restore essential highway traffic, minimize the extent of damage, or protect the remaining facilities, which are incurred in the first 180 days, starting with the date of the letter of intent for the disaster, receive100 percent Federal share reimbursement. All permanent restoration costs and any costs incurred after the first 180 days are to be reimbursed at the normal pro rata share.

For eligible basin flooding disasters, ER funding will generally be limited to grade raises of highways and bridges. However, if within the basin, Federal-aid routes have concurrently suffered traditional ER eligible damage, such as culvert washouts or embankment failures, repair of this traditional damage need not be handled as a separate ER disaster but may be included as incidental to the basin flooding disaster.

Although basin flooding is considered as an ER disaster, it is expected that ER requests be limited to those truly unusual events that meet the criteria outlined above. It is not the intent of the ER program to raise the grades in dips or other low-spots along Federal-aid routes that suffer chronic undulation flooding problems. Rather, these latter situations should more appropriately be corrected using regular Federal-aid funding sources.

7. Slides

The removal of rock and mud slides from a Federal-aid highway is eligible when the slide is either associated with the overall natural disaster or by itself qualifies as an eligible natural disaster. In both cases, its correction to provide a safe roadway is eligible. If found cost-effective, ER funds can be used to stabilize slide areas to protect a highway facility from future disaster damage (see further discussion of betterments at the end of this chapter). Such stabilization is considered to be an ER-eligible betterment. Also, relocation of the road rather than slide correction is an eligible betterment if found to be cost-effective and accompanied by proper documentation.

8. Work on Active Construction Projects

Most States require the contractor to take all necessary precautions to protect Federal-aid projects from damage, including ER-funded projects still under construction or practically completed, but not yet accepted by the State. Before considering the ER eligibility of work on an active construction project, it must clearly be established that such work is not the responsibility of the contractor. The contract documents may specify the responsibility for damages that are caused by a disaster during the construction stage of a project. Generally, project elements that are in place and accepted by the State prior to the disaster would be eligible for ER funding. In some cases, the contract documents may provide for Federal-aid funding eligibility as an addendum to the contract. Such a provision may allow for reimbursement with other Federal-aid funds, but not necessarily ER funds. Damages to materials stockpiles and contractor equipment are not eligible for ER funding.

9. Toll Facilities

ER funds may participate in repair of Federal-aid highway toll facilities when the provisions of section 129 of 23 USC. are met. For example, if a toll facility is on a National Highway System (NHS) route but does not have a section 129 toll agreement in place, then the designation of the route alone does not make this eligible for ER funding. To be eligible for ER funding, the highway must be a Federal-aid highway in accordance with the definition in 23 USC. 101(a)(5) which states that "[t]he term 'Federal-aid highway' means a highway eligible for assistance under this chapter other than a highway classified as a local road or rural minor collector." Since a toll highway that does not have an active section 129 toll agreement in place is not eligible for assistance under title 23, such a roadway cannot receive ER funding. However, if an existing toll facility does not have an executed toll agreement with the FHWA at the time of the disaster, a toll agreement may be executed after the disaster to qualify the facility for repairs. Loss of toll revenue is not eligible for reimbursement.

10. Traffic Control Devices

ER funds can participate in the cost of the repair or replacement of traffic control devices including traffic signal systems, directional and informational signs, and railroad-highway crossing warning devices, if the damage and associated repair or replacement costs can be shown to exceed heavy maintenance.

11. Landscaping

Replacement of damaged highway landscaping is eligible if it represents incidental cost associated with otherwise eligible damage. For example, following a wildfire, erosion control (including hydroseeding) of damaged areas within the highway right-of-way is an eligible activity. The eligibility of erosion control and hydroseeding outside the right-of-way should be considered on a case-by-case basis. ER participation in erosion control outside the right-of-way may be economically justified if an analysis demonstrates that the cost of such repairs will result in a net savings to the ER program should a future disaster affect the site. Additional discussion of eligible repairs outside the highway right-of-way can be found in item 16 below.

12. Roadside Appurtenances

The repair or necessary replacement of damaged or destroyed guardrail, bridge rail, impact attenuators, right-of-way fences, etc., is eligible if such damage is extensive and not limited to a few isolated cases under each category.

13. Timber and Debris Removal

Only debris that is deposited as a direct result of a disaster is eligible for ER funding. Any debris that was not deposited as a direct result of a disaster, such as debris collected and placed by the roadway from an adjacent property-owner, is not eligible for ER funding. However, such debris may be eligible under FEMA's Public Assistance Program. Debris removal is considered emergency repair only to the extent that it is needed to minimize damage, protect facilities, or restore essential traffic. Some Divisions provide guidance on debris removal to facilitate distinguishing between debris that is deposited by the event (ER-eligible debris) and debris that may have been added to the roadside following initial debris pickup (FEMA-eligible debris). Divisions are encouraged to work with FEMA to establish such guidance prior to an event.

The Federal share for the emergency repair portion of debris removal can be at 100 percent within the first 180 days of the event. Typically, the limits of debris removal at 100 percent within the cross-section should be to the outside edge of the shoulders and can include the removal of debris that is considered to be a safety hazard (fixed objects) within the limits of the clear zone. Debris removal outside these limits during the first 180 days and beyond is at the normal pro rata share for the affected Federal-aid highway.

The cost of stockpiling and disposing of debris at adjacent sites, as well as the cost of removing marketable timber from the acceptable clearing limits and transporting to adjacent stockpile sites, is eligible for ER participation. However, hauling costs to sawmills or to other locations beyond the general proximity of the damaged highway are not eligible. The clearing limits for debris, including downed timber, normally, should include the traveled way, cut and fill slopes and any additional clearing required to assure the full functioning of the pavement, drainage ditches, and structures, including the clear zone for safety. Clearing of the remainder of the full right-of-way is the responsibility of the agency having jurisdiction. Cut sections should be cleared to the safe distance that will assure that no debris will cause roadway slope erosion or will roll down to clog ditches or endanger traffic on the pavements and shoulders. The timber and debris removal operations should conform to the standards of safety for that particular route.

In the case of normal medians, the necessary cleanup of downed timber and debris is eligible. Where directional roadways or "divided highways" are widely separated because of terrain or for aesthetic reasons, the cleanup of the entire median would not be eligible. Each directional roadway should be treated as a separate roadway, including cut and fill slopes, and handled as described in the above paragraph.

Snow and ice removal are not eligible as debris removal.

14. Transportation System Management (TSM) Strategies

TSM strategies to monitor and control traffic and to manage transportation on streets and highways during and immediately following a disaster to restore traffic are eligible for ER funding.

The elements of the TSM plan may include highway advisory radio, closed circuit television, video image process surveillance, installation of changeable message signs, a public awareness program, etc.

ER funds are eligible to provide emergency traffic management services by the police during and immediately following a disaster under the following conditions: Such traffic services by the police must 1) directly relate to the disaster; and 2) represent added costs above and beyond costs related to normal, day-to-day responsibilities, i.e., overtime costs or additional shift costs. The added costs should be itemized and documented to support any use of ER funds for this activity. Lost ticket revenue due to redirection of traffic enforcement police services resulting from an emergency is not an eligible ER expense.

15. Projects and Project Features Resulting from the NEPA Process

An ER repair project developed in accordance with the National Environmental Policy Act (NEPA) process may incorporate added features to mitigate environmental impacts to such resources or sites as wetlands, noise-sensitive land uses, endangered species, etc.

These added mitigation features, by themselves, are not considered "betterments" for the purposes of the ER program since they do not protect the highway from future disasters or change the function or character of the highway facility from what existed prior to the disaster or catastrophic failure.

This is illustrated by the following example:

Environmental Mitigation Feature Eligible for ER Funding

In repairing a damaged bridge, some of the construction activity will take place in or closely adjacent to the existing stream bed. This particular stream serves as the spawning grounds for an endangered species of fish. The contractor would be required to use special construction techniques that minimize disruption of the streambed, and special pooling areas for the fish will need to be built in the stream adjacent to the bridge. The special construction techniques and the added stream features are necessary mitigation measures to implement the repair project, and they would be eligible for ER funding.

Eligibility determinations for environmental mitigation measures can apply a general rule-of-thumb: if the mitigation measure is related to an ER-eligible betterment, the mitigation measure is also eligible.

For example, if a roadway grade raise to protect a facility from future flooding has been economically justified for ER funding, then a mitigation feature associated with the grade raise, such as possible wetland restoration, would qualify for ER funding. Conversely, if a "betterment" is not justified for ER funding, then any added mitigation features related to the betterment would likewise not be eligible for ER funding but instead should be funded from regular apportioned Federal-aid highway funds.

For ER replacement projects, the NEPA decision process can also determine project location, potentially including a completely new location. Further, the mere fact that something old is being replaced with something new, or that the new facility is built to current design standards, does not restrict ER funding.

In general, if it has been determined that ER funding may fully participate in the replacement project (see the various scenarios discussed on page 31), and if the replacement project is of comparable function and character to what existed prior to the disaster or catastrophic failure, then ER funding may fully participate in the replacement project.

As noted, the NEPA process may well determine the location of the replacement project. In the case where it is neither practical nor feasible to replace a destroyed facility in-kind at its existing location, the alternative facility developed in conformance with the NEPA process is eligible for ER funding, as illustrated by this example:

Replacement Facility on New Location Eligible for ER Funding

A roadway was permanently submerged by water backing up behind a naturally created dam, and replacement of the inundated highway facility at its existing location is neither practical nor feasible. Through the NEPA process, various alternative locations for a replacement facility were evaluated. A recommended alternative emerged from the process. This recommended alternative was of comparable function and character to the damaged facility (i.e., same number of lanes, same degree of access control), and, accordingly, was eligible for ER funding.

The NEPA process may also require that a replacement project include additional project features to mitigate impacts of the project.

These added mitigation features are eligible for ER funding if the replacement project itself is eligible for ER funding. An example follows:

Environmental Mitigation Feature Added to Replacement Facility Eligible for ER Funding

A replacement facility was predicted to result in increased noise impacts on adjacent residential property, and the NEPA noise impact evaluation determined that noise walls must be included in the final project. Although the damaged facility did not have existing noise walls, this added feature, as an environmental commitment, became an integral part of the replacement project in compliance with NEPA and FHWA's Title 23 highway noise impact assessment and mitigation requirements. Further, since a noise wall would not protect the highway from future disaster damage or change the function or character of the highway facility, it did not have to be justified as a "betterment" under the ER program. As a result, this noise wall was eligible for ER funding.

The above discussion on replacement facilities and use of ER funding, in general, reflects those situations where ER funding is not capped. However, if replacement facilities are being advanced where ER funding participation is capped, this same limit on ER funding would apply to the proposed replacement facility including any environmental mitigation features required as a result of developing the replacement facility in accordance with NEPA.

16. Outside of the Highway Right-of-Way

Generally, damage repair activities outside the highway-right-of-way are not eligible for ER funding. The exception would be work associated with stream channels adjacent to a highway. Normally, projects associated with channel work (riprap, bank protection, etc.) that require right-of -way purchases and/or easements outside the right-of-way are not eligible. The fact that the agency responsible for channel maintenance does not have funds to finance the repair and protection work is not an acceptable justification for ER fund assistance. In situations involving requests for participation in erosion control and bank protection outside the highwayright-of way, the following items must be verified by the highway agency to obtain eligibility:

- The work is directly related to protection of the highway facility;
- The work is not eligible for funds from another agency;
- No other agency has the responsibility for such work;
- The applicant agrees to accept the future maintenance of all work performed.

When work of this type is proposed, the project documents should include a letter from the State agency showing how all of the above four items have been or will be satisfied.

17. Administrative Expenses

a. Regular and Extra Employees

Regular salaries and overtime salaries and wages of all regular and extra employees of the applicant directly engaged in the performance of work on ER projects are eligible for reimbursement. Timekeeping procedures should facilitate allocating employees' time to projects, and/or other activities, each day on an hourly basis. A timekeeping document, such as time slip, time and attendance report, or time book, must be available for examination by audit personnel to support direct labor costs claimed on any ER project. A responsible employee having knowledge that the time distribution is accurately reported should sign the document.

b. Payroll Additives

Usually referred to as labor surcharge, a set percentage over and above the total direct labor costs charged to a project is eligible for ER participation. This surcharge covers costs of various types of leave allowances, industrial accident insurance coverage, and other employee benefits. The allowable percentage rates will normally vary from year to year. Also, employee benefits allowed by one applicant may differ widely from those allowed by another. Therefore, the records used in developing percentage rates should be preserved under suitable control conditions to assure availability for examination when requested. The acceptable percentage rate may be applied only to direct labor costs.

18. Supplies and Materials

Engineering and general office supplies of an expendable nature, charged from stock or purchased for a particular project, and properly identified on the stock-issue slip or vendor's invoice with the project charged, are considered eligible for participation. In the case of issues from stock, verifiable evidence to assure validity of the prices charged must be available for examination if requested. Many classes of materials required for ER projects are purchased for a particular project. In such cases, the costs are eligible for participation, provided the project on which the materials are used is properly identified on the supplier's invoice. The cost of materials issued from stock warehouses or agency-produced or purchased stockpiles for use on an ER project must be properly supported. The records supporting the prices charged should be available for audit when requested. Also, a responsible employee having knowledge that the supplies or materials have actually been incorporated into the project should sign the source document (stock issue slip or supplier's invoice).

19. Equipment

The use of applicant-owned equipment or equipment owned by, and rented from, another public entity, or rented equipment from commercial sources, which is necessary for the prosecution of work properly authorized under an ER project, is eligible for participation. Such use should be charged on a per hour, per shift, per mile, etc., basis. The rental cost of such equipment may include such items as normal operation (gasoline, fuel oil, lubricants, antifreeze, etc.); repair (major and minor), etc.; and depreciation or replacement expenses. Costs in excess of normal depreciation to cover the increased costs of replacement equipment are to be excluded from equipment rental rates applied to ER projects. The rates per unit for applicant-owned equipment must be properly supported by verifiable analyses covering specific sizes and types of equipment charged. Lacking such documented cost analyses by the applicant, ER funds will participate in the equipment rental costs on the basis of rates developed by the State DOT and approved by the Division Administrator. Each applicant should use either the FHWA approved State DOT rates or its own established rates, but not a combination of both. The required document action to support equipment rental charges is an acceptable time and equipment use report, signed by a responsible employee signifying actual use of the equipment on the project designated.

Reasonable costs of equipment rented from commercial sources for use on ER projects are eligible for ER participation. The extent of "reasonableness" will be determined by the Division Administrator as consistent with the current edition of the *Associated Equipment Distributors Manual* or rates charged by a recognized rental agency. A commercial invoice, signed by a responsible employee signifying actual use of the equipment on the project designated, is required documentation to support equipment rental charges. The supporting document must indicate the period of use applicable to an identified ER project. The applicant may also claim operating costs provided that the rental costs do not include operating cost. Equipment "move in" and "move out" costs may also be considered eligible.

20. Catastrophic Failure from an External Cause

A catastrophic failure from an external cause is an eligible ER expense. A catastrophic failure from an external cause is defined as the sudden failure of a major element or segment of the highway system due to an external cause. The failure must not be primarily attributable to gradual and progressive deterioration or lack of proper maintenance. The closure of a facility because of imminent danger of collapse is not in itself a sudden failure.

Examples of such disasters include a barge striking a bridge pier causing the sudden collapse of the structure, a truck crash resulting in a fire that damages the roadway, and acts of terrorism that result in structural damage to the highway facility.

Prompt and diligent efforts shall be made by the State to recover repair costs from the legally responsible parties to reduce the project costs particularly where catastrophic damages are caused by ships, barge tows, highway vehicles, or vehicles with illegal loads or where damage is increased by improperly controlled objects or events. [see 23 CFR 668.105(f)]

Any compensation for damages or insurance proceeds including interest recovered by the State or political subdivision or by a toll authority for repair of the highway facility must be used upon receipt to reduce ER fund liability on the project. [see 23 CFR 668.105(e)]

Funding is to be credited to the ER project for which compensation is recovered.

FHWA's share of the recovery amount should be proportionate to the ER funding percentage that participated in the project. Any resulting unneeded balance of ER funding on that project will then be withdrawn from the State and made available for other nationwide ER needs.

C. Ineligible Items

ER funds are not intended to cover all damage repair costs. Only that repair work which exceeds heavy maintenance, is extraordinary, and will restore pre-disaster service is eligible. Incidental costs resulting from a disaster, such as project delay costs or lost toll revenues, are not an eligible ER expense. A more detailed discussion of repair activities that are not eligible for ER funds follows.

1. Heavy Maintenance

When a disaster has caused damage requiring heavy maintenance or work frequently performed by the applicant's maintenance crews, repairs are not eligible. Heavy maintenance is usually performed by highway agencies to repair damage normally expected from seasonal and occasionally unusual natural conditions or occurrences. It includes work at a site required as a direct result of a disaster that can reasonably be accommodated by a State or local road authority's maintenance, emergency, or contingency program. Examples include work necessary to repair minor damage due to eroded shoulders, filled ditches and culverts, pavement settlement, mud and debris deposits, slope sloughing, and slip-outs in cut or fill slopes.

It is not the intent of the ER program to relieve an agency of its maintenance responsibility simply because a storm of unusual character and extent causes serious damage to roads and to public and private property.

2. Damage Estimate under $5,000 per Site

A minimum $5,000 in repair cost per site is used as a guideline for a site to be ER eligible. This minimum guideline dollar value is suggested for national consistency but, if circumstances warrant and the State has requested a different figure, the Division Administrator may allow a different minimum amount. This dollar threshold criterion is used to distinguish repairs that are maintenance activities from an ER program responsibility. Generally, a site is an individual location where damage has occurred. However, a site could include several adjoining locations where *similar* damage, related to the same cause, has occurred. For example, where flooding waters in a drainage course that runs along or continually crosses under a highway has caused damage at several locations within a reasonable distance of each other, it would be acceptable to package these together as a single site.

Another exception could involve damage to traffic signs which has occurred at several locations on an area wide basis. In this case it would be acceptable to group these locations by route or jurisdiction, with the accumulated amount considered a site.

Overall, aggregating locations together to form a site should be done with care, bearing in mind that the intent of the ER program is not to pay for highway damage repairs that a transportation agency would normally perform as heavy maintenance. For this reason, grouping damages to form a site based solely on a political subdivision (i.e., county or city boundaries) should not be accepted.

3. Traffic Damage

Repair of surface damage caused by traffic, whether or not the damage was aggravated by saturated subgrade or inundation, except as noted under Eligible Item B4 above, is not eligible for ER funds.

4. Frost Heaving

Damage to roadway subbase and base materials due to inundation or freezing and resultant frost heaves, even if the roadbed has been saturated by the floodwaters, is not eligible for repair with ER funds. Surface roadway damage such as potholes resulting from such conditions are not eligible for ER funding.

5. Applicant-Owned Material

Replenishment or replacement of destroyed or damaged stockpiles of materials for both maintenance and construction that have not been incorporated into the highway is not eligible under the ER program.

Repair of damage to or replacement of an applicant's or contractor's facilities, such as maintenance sheds or equipment, is not eligible for ER funds, for example:

> *The applicant is in the process of widening a bridge as a maintenance betterment project.*
>
> Precast concrete beams have been delivered to the job site but have not been incorporated into the structure. Rapid runoff of floodwaters destroys the existing bridge and the precast beams. Cost of repair of the existing bridge is eligible. However, since the precast beams were not a part of the existing bridge at the time of its destruction, cost of replacing the stockpiled beams is not eligible.

6. Erosion Damage

Repair of minor erosion damage due primarily or wholly to rainfall and resulting from surface saturation of slopes and embankments, rather than from floodwaters, is not eligible. Even though the work may be a little more extensive than usual, it is considered heavy maintenance.

7. Prior Scheduled Work

Permanent repair or replacement of deficient bridges scheduled for replacement with other funds damaged during a disaster is not eligible for ER funds, and should be funded as originally intended. The ER funds may participate in emergency repairs to restore essential traffic in such cases.

A project is considered scheduled if the "construction phase" of a replacement structure is included in the FHWA approved Statewide Transportation Improvement Program (STIP). As used in this section, the term "construction phase" refers to the physical construction separate from any other identified phases in the STIP such as planning, design, or right-of-way phases.

Inclusion of bridge replacement work in a city or local agency capital improvement plan is viewed by FHWA as prior scheduled work and therefore is not eligible for ER funding.

In this case, the city or local agency is that agency responsible for capital improvement program funding decisions for that jurisdiction and such a plan is officially recognized for that purpose.

8. Mine/Underground Subsidence

Where a highway passes over an underground mine and a section of highway is closed down due to actual mine subsidence, repair work on the section of highway exhibiting physical damage due to subsidence is eligible for ER funding.

However, repair work to prevent additional physical damage to adjacent sections of the highway over the mine is not eligible. Further, in situations where in-bed stream mining is found to be contributing to erosion (where a highway passes over or is adjacent to the stream), associated repair work is considered ineligible.

9. Snow and/or Ice Removal

Snow and/or ice removal is not eligible for ER funding as snow and/or ice is viewed as a relatively short term event not causing physical damage to a highway. Although ER funds may not be used for snow removal, the FEMA may allow funding available through its assistance program to pay for a limited amount of snow clearance on Federal-aid highways during extreme snowfalls.

10. Emergency Transportation Services/First Responders

Emergency medical transportation services, such as ambulances or helicopters, are not eligible for ER funding. Activities associated with first responders to a disaster such as emergency police services, firefighting, emergency medical, and any other rescue-related activities including the use of divers for rescue operations, are not an eligible ER activity.

11. Mitigation/Preventive Work/Evacuation Prior to a Disaster

Preventive work to avoid damage to a highway facility in anticipation of a disaster is not eligible for ER funding. For example, work to prevent scour at bridge sites in anticipation of extremely high rainfall and potential flooding is not eligible for ER funding.

Costs associated with evacuation activities prior to a disaster such as traffic control and police and traffic management activities associated with evacuation of an area prior to the actual occurrence of an event, such as a hurricane, are not eligible for ER funding.

12. Catastrophic Failure from an Internal Cause

Not all catastrophic failures are ER eligible. An ER eligible catastrophic failure must be a sudden failure of a major element, or segment of the highway, due to an external cause. ER funds do not apply to catastrophic failures from an internal cause or source, e.g., gradual and progressive deterioration or lack of proper maintenance. Closure of a facility because of imminent danger of collapse is not, in itself, a sudden failure and therefore is not eligible for ER funding.

13. Radiological Contamination

Acts of terrorism may result in radiological contamination of a Federal-aid highway facility where no associated structural damage has occurred. Under such a scenario, radiological contamination may render the bridge and/or the surrounding area temporarily or even permanently inaccessible. ER funds cannot be used solely for the purpose of radiological decontamination of a highway infrastructure where there is no incidental structural damage.

If the event resulted in structural damages that met the minimum ER event threshold, then decontamination associated with the repair or replacement of the damaged facility could be considered an eligible ER expense.

14. Transit Operation and Maintenance Costs

The participation of ER funds for the operation of substitute emergency transit services is not permitted under the current ER statute. ER funds may participate in the operation and maintenance costs of ferryboats providing temporary substitute highway traffic service around a damaged highway facility (see Chapter II, Section B-3, *Detours*).

D. Emergency Repairs vs. Permanent Repairs

Both emergency repairs and permanent repairs are eligible for ER funds. Emergency repairs are repairs made during and immediately following a disaster to restore essential traffic, to minimize the extent of damage, or to protect the remaining facilities. Permanent repairs are repairs undertaken, normally after emergency repairs have been completed, to restore the highway to its pre-disaster condition. It is important to distinguish between emergency and permanent repairs because emergency repairs accomplished during the first 180 days after the occurrence of the disaster are funded at a higher Federal share.

1. Emergency Repairs

a. General

Emergency repairs are repairs made during and immediately following a disaster to restore essential traffic, to minimize the extent of damage, or to protect the remaining facilities [see 23 USC. 120(e)]. The State and local transportation agencies are empowered to begin emergency repairs immediately without prior FHWA authorization. Properly documented costs will later be reimbursed once the FHWA Division Administrator makes a finding that the disaster is eligible for ER funding. Typical examples of emergency repairs are:

- Regrading of roadway surfaces, roadway fills, and embankments
- Debris removal
- Erection and removal of barricades and detour signs, flagging and pilot cars during the emergency period, and placement of riprap around piers and bridge abutments to relieve severe on-going scour action
- Dynamiting and other removal of drift piling up on bridges, including rental of boats
- Placement of riprap on the downstream slopes of approach fills to prevent scour during overtopping of the fill
- Removal of slides
- Construction of temporary roadway connections (detours)
- Erection of temporary detour bridges
- Replacement of approach fills
- Use of ferryboats to provide temporary substitute highway traffic service

Any such work may subsequently be included in an ER program of projects, which, when submitted for approval, should include both emergency repairs and any permanent restoration work performed coincidentally with the emergency repairs.

b. Intent of Emergency Repairs

The intent of emergency repairs is to permit the immediate performance of work to restore essential traffic that cannot wait for a finding of eligibility and programming of a project. Emergency repair work should be accomplished in a manner that will reduce additional work required for permanent repairs.

The Division Administrator's concurrence on the need for emergency work conveyed in the Division's acknowledgment to the State's letter of intent does not authorize the agency to proceed with permanent restoration work on damaged highways unless the restoration is more economical or practical to perform as an associated part of the emergency operation. This incidental restoration can be performed simultaneously with the emergency work provided it is properly covered in the damage inspection report. If such work has been accomplished prior to the site damage review, retroactive approval may be given when circumstances warrant.

There may be situations where immediate completion of the permanent work is the most economical and feasible operation to quickly restore essential traffic. An example would be the washout of a bridge and its approaches where construction of a temporary detour would be both costly and time-consuming, but the agency involved has precast concrete girders readily available that could be used at the site. In such a case immediate construction of the permanent structure and approaches could be accomplished at the discretion of the State, and *approval* by FHWA of the permanent repair would be handled as an emergency repair. However, such permanent repair work is still to be reimbursed at the normal pro rata share for that facility even if the permanent repair is performed as part of the emergency repair work.

The use of ER funds for emergency repairs to roadways is normally limited to the amount necessary to bring the washed-out fills and slip-outs back to grade with a gravel surface. In most cases these emergency repairs to roadways are not constructed to true line and grade. They usually follow the terrain and are constructed in the easiest and most expedient manner. The repair to the road, nevertheless, should be sufficient to permit traffic to travel over it safely. Where routes handle heavy traffic, an appropriate type of bituminous surface as a temporary repair will be eligible for short sections of roadway.

The placement of the final high-type pavement is normally considered permanent repair. If performed as emergency work, such paving must have FHWA concurrence to be eligible for Federal participation. A determination of whether or not the work should be performed as emergency or permanent should take into account traffic characteristics, remoteness of the site, traffic control requirements, and socioeconomic factors.

c. Federal Share (180-Day Period)

Emergency repairs accomplished within the first 180 days of the disaster occurrence to restore essential traffic, minimize the extent of damage, or protect the remaining facilities may be reimbursed at 100 percent Federal share.

The 180-day time period for 100 percent Federal share is established by 23 USC. 120(e), and the FHWA has no authority to change the time period. It is important to give careful consideration in deciding the beginning of the 180-day time period. The 180-day time period is intended to start on the initial day of the disaster occurrence within a particular State. The starting date of a disaster is to be applied on a statewide basis.

As previously noted in Chapter I, Section B, *Program Overview*, permanent repair work is not to be considered emergency repair work for the purpose of establishing the eligible Federal share, and can only reimbursed at 100% if special legislation allows.

2. Permanent Repairs

a. General

Permanent restoration work is not eligible if performed prior to authorization by the FHWA, unless it is determined to be more economical or practical to perform such work as an associated part of the emergency repairs; documentation of this determination is required. Permanent repairs should be administered using normal Federal-aid contracting procedures although use of streamlined procedures is encouraged (e.g., A + B bidding, reduced advertising period, etc.). Such repairs may involve one or more of the following categories of work.

b. Restoration-in-Kind

The ER program provides for the repair and restoration of highway facilities to pre-disaster conditions. Restoration in kind is therefore the predominant type of repair accomplished with ER funds. Any additional protective features or changes to the function or character from that of the pre-disaster facility are generally not eligible for ER funding unless justified on the basis of economy of construction, prevention of future recurring damage, or technical feasibility.

c. Replacement-in-Kind

(1) Existing Location

Where extensive damage has occurred, ER funds may be used for replacement-in-kind as the proper solution. If a facility has been damaged to the extent that restoration to its pre-disaster condition is not technically or economically feasible, a replacement facility is appropriate. Replacement facilities should be constructed to meet current design standards.

ER participation in a replacement roadway will be limited to the costs of a roadway built to current design standards, and of comparable capacity, (e.g., number of lanes), and character, (e.g., surfacing type, access control, rural/urban section), of the destroyed facility. ER reimbursement for a replacement bridge will be the cost of a new bridge built to current design standards for the type and volume of traffic the bridge will carry during its design life.

ER participation may be prorated. Where the replacement project exceeds ER eligibility limitations, the ER share of project cost will be limited to the estimated cost of the ER eligible replacement roadway or bridge.

(2) New Location

Where relocation is necessary, each case must be considered carefully to determine what part of the relocation is justified for construction with the participation of ER funds. This matter is discussed further in the sections covering environmental considerations in chapter VI and betterments in this chapter.

d. Deficient Bridges

This category includes structural conditions only. It does not consider waterway opening, functional obsolescence or serviceability, etc. Permanent repair of a structurally deficient damaged bridge is eligible for ER funding provided a replacement bridge is not under construction or the bridge is not scheduled for replacement (in the FHWA approved STIP) with other Federal funds. Inclusion of bridge replacement work in a city or local agency capital improvement plan is viewed by FHWA as prior scheduled work and therefore is not eligible for ER funding (see Chapter II, Section C-7, *Prior Scheduled Work*). The intent is to ensure that ER funds do not replace other Federal funds that would have otherwise been used to construct a replacement facility.

The following represent two common situations:

(1) Bridge Is Damaged But Is Repairable

ER funds may participate in:

- Reasonable emergency repair to restore travel
- Permanent repair of disaster damage if bridge will be structurally safe upon completion of disaster damage repair (meaning that repair of disaster damage corrects structural deficiency)
- Permanent repair of disaster damage if other funds are used to simultaneously correct the structural deficiencies (meaning that repair of disaster damage will not correct structural deficiency)
- No permanent repair if bridge is scheduled for replacement

(2) Bridge Is Destroyed or Repair Is Not Feasible.

ER funds may participate in:

- Reasonable emergency repairs to restore traffic
- New comparable replacement structure to current design standards and to accommodate design-year traffic volume if bridge is not scheduled for replacement
- No permanent repair if bridge is scheduled for replacement in the current FHWA approved STIP or local capital improvement plan or a replacement bridge is under construction.

e. Replacement of Culverts

Upgrading of culverts to current standards must be specifically related to eligible disaster damage repair.

- Damaged culverts are eligible for repair in kind.
- Destroyed culverts are eligible for replacement to current standards.
- Wholesale upgrading of deficient culverts on an area or route basis is not eligible. Eliminating a recurring annual maintenance problem with ER funds, based on the occurrence of a disaster, is not normally within the scope of the ER program.

f. Betterments

Betterments, for the purposes of the ER program, are defined as (i) added protective features, such as the rebuilding of roadways at a higher elevation or the lengthening of bridges, or (ii) changes which modify the function or character of a highway facility from what existed prior to the disaster or catastrophic failure, such as additional lanes or added access control.

(i) Betterments Involving Added Protective Features:

This category of betterments involves adding features to highway facilities that help protect these facilities from possible future damage from ER-eligible disasters or catastrophic failures. Examples include:

- Raising roadway grades
- Relocating roadways to higher ground or away from slide prone areas
- Stabilizing slide areas
- Stabilizing slopes
- Installing riprap
- Lengthening or raising bridges to increase waterway openings
- Deepening channels
- Increasing the size or number of drainage structures
- Replacing culverts with bridges
- Installing seismic retrofits on bridges
- Adding scour protection at bridges
- Adding spur dikes

If a betterment involving an added protective feature is included in an ER repair project, the betterment may be considered eligible for ER funding if it can be economically justified based on an analysis of its cost versus projected savings in costs to the ER program should future ER-eligible disasters occur within the normal design year for the basic repair work. This cost/benefit analysis focuses solely on benefits resulting from estimated savings in future recurring repair costs under the ER program. The analysis does not include other factors typically included in highway benefit/cost evaluations, such as traffic delays costs, added user costs, motorist safety, economic impacts, etc.

The following example illustrates this type of betterment and the economic analysis that would be applied in determining whether the betterment would be eligible for ER funding.

Added Feature Justified for ER Funding

Floodwaters have damaged a 1,400-foot section of roadway with both surfacing and fill material being washed away. The site can be repaired for $150,000. However, this section of roadway was similarly damaged during an ER event several years ago, and it appears likely to be damaged again in the future. The State proposes to add additional culverts to increase the drainage capacity under the roadway, thus substantially reducing the chances that this type of flooding in the future will damage this section of roadway. Adding these culverts is estimated to increase the cost of the repair project by an additional $100,000. Spending an additional $100,000 for the added features could save one or more future ER repair costs that could easily exceed the cost of the added features. In this case, the added features are economically justified because of the potential to save future costs under the ER program, and, consequently, the added features are eligible for ER funding.

In the previous example, the betterment is considered eligible for ER funding. However, in many instances betterments will fail to meet the economic justification test for use of ER funding. If ER funding cannot be provided for a betterment, this does not mean that the betterment should necessarily be excluded from the ER repair project. If a betterment provides considerable benefit when other factors are considered, the State is encouraged to use regular apportioned Federal-aid highway funds, as appropriate, to fund a betterment. This point can be illustrated by this example:

Added Feature Not Justified for ER Funding

Flooding waters covered a 1,000-foot section of a heavily traveled highway. Traffic must be detoured to other routes, causing extra travel and considerable delays. The highway was unusable for 3 weeks until the floodwaters receded. At that time, it was found that the flooding caused only limited damage to the highway, which can be repaired for $75,000. However, the extra travel distances and delays during the 3-week period the highway was closed imposed an estimated $1,100,000 in additional costs on the motorists and community. The State proposed to raise the grade of the highway by 5 feet, thus placing the roadway at an elevation where a similar flood would not inundate the roadway. The estimated cost of raising the grade for this section of highway was $950,000. Although the cost of the grade raise could easily pay for itself in terms of potential savings in travel and delay costs to the motorists and community should another similar flood occur, the cost is substantially greater than the potential repair costs that would be eligible under the ER program. In this case, the added feature would not be considered to be economically justified for ER funding. Instead, the more appropriate Federal highway funding source would be regular apportioned Federal-aid highway funding.

There are limited situations where added features require no further economic justification as betterments. One situation applies to reasonable grade raises associated with basin flooding. Another involves repairs of features, such as bridges, that may require permits or approvals from other entities. If these other entities routinely require added features as standard industry practice on other projects of similar nature to the ER project, then these added features can be included on the ER project without further justification as betterments.

(ii) Betterments Involving Changes to the Function or Character of the Highway:

This category of betterments involves making changes to the function or character of the facility from what existed prior to the disaster or catastrophic failure. Examples include:

- Adding lanes
- Upgrading surfaces, such as from gravel to paved
- Improving access control
- Adding grade separations
- Changing from rural to urban cross-section

In general, betterments that change the function or character of a facility do not qualify for ER funding. One exception established under 23 USC. 120(e) allows ER funding participation in replacement bridge facilities to accommodate traffic volumes over the design life of the bridge, thus potentially allowing ER funding for added lanes on bridges. This type of betterment, and eligibility of ER funding, is illustrated in the following examples:

Change in Function/Character Eligible for ER Funding

A 2-lane bridge, on a heavily traveled urban route, is destroyed by floodwaters. The State proposes to build a replacement bridge at the same site. The design year traffic (20 year traffic projection) for the bridge shows that a 4-lane facility will be needed to accommodate this future traffic. Accordingly, the State proposes that the new bridge have 4 lanes, a significant improvement in capacity above the destroyed bridge. As a result of the statutory provision in 23 USC. 120(e), the added lanes on the new bridge are eligible for ER funding.

Change in Function/Character Not Eligible for ER Funding

A 2,000-foot section of a 2-lane heavily traveled urban route was seriously damaged by floodwaters. The State proposed that repair of the section include widening to four lanes to accommodate future traffic needs since at some future time the State will need to widen the route anyway. This added capacity is viewed as changing the function and character of the route, and the costs of the additional lanes are not eligible for ER funding. However, since widening the route at the same time the repair work is done may be in the public interest by providing some savings in costs and by avoiding later disruptions to traffic, a State could decide to use its regular apportioned Federal-aid highway funds for the added costs of the extra lanes.

Repair Activities that Are Not Betterments:

The dictionary defines the term betterment as an improvement that adds value. The matter of adding value is subject to wide interpretation. As noted in the previous discussion, for the purposes of the ER program, the term betterment applies specifically to added protective features or changes made to the function or character of the facility. With this in mind, there are certain classes of ER funded repair activities that are not considered betterments.

These are:

- Replacement of older features or facilities with new ones--The mere fact that a damaged highway feature or facility is replaced with something new that may extend the service life of the facility, in and of itself is *not* a betterment.
- Incorporation of current design standards--Repaired facilities may be built to current design standards, which could result in improved or added features that do not change the function or character of the facility. For example, a repaired length of roadway may have wider lanes or shoulders and additional roadside safety hardware that result from following current design standards. These features are not betterments.
- Replacement in-kind on existing location not practical or feasible -- On rare occasions, it is neither practical nor feasible to replace a damaged highway facility in-kind on its existing location. An alternative selected through the NEPA process, if of comparable function and character to the destroyed facility, is eligible for ER

funding. This is *not* a betterment. (See the following discussion in this chapter on replacement facilities for more information on this special situation.)
- Additional required features resulting from the NEPA process -- ER projects may include additional required features as an outcome of the project being developed in accordance with the NEPA process. These features are eligible for ER funding. These are *not* betterments.

Replacement Facilities and Betterments:

Restoration in-kind is the predominant form of repair activity under the ER program. However, at times restoration is not possible, and replacement is necessary, or a State may choose to replace a facility rather than repairing it. The extent of ER participation varies depending on the circumstances involved. Various scenarios are discussed below.

In a first replacement scenario, a highway facility was seriously damaged; however, inspection verified that repair and restoration of the existing facility was possible. However, the State preferred to replace the existing facility with a new or alternative facility. In this case, ER funding limited to the ER program share of the estimated cost to repair the existing facility can be applied to a new or alternative replacement facility. Regular apportioned Federal-aid highway funds may be used to fund project costs above the amount eligible for ER funding.

In a second replacement scenario, a highway facility was seriously damaged, and inspection confirmed that it was not repairable. The State decided to replace the existing facility with an in-kind replacement facility (of comparable function and character to the damaged facility) on the existing location. In this case, ER funding may participate in the total cost of the replacement facility.

In a third replacement scenario, inspections confirmed that a seriously damaged highway was not repairable. Although building a replacement facility at the location of the existing facility was found to be feasible, the State chose to replace the existing facility with a new facility, of comparable function and character, on a new location. In this case, ER funding for the new facility is limited (capped) to the ER program share of the estimated cost to replace the facility in-kind at its existing location. In this scenario, it may be possible to utilize regular apportioned Federal-aid highway funds to fund project costs above the amount eligible for ER funding.

In a fourth replacement scenario, inspection has again confirmed that a seriously damaged highway is not repairable. Additionally, because of the unique circumstances involved, in-kind replacement at the existing location was determined neither practical nor feasible. Consequently, an alternative replacement facility on a new location was developed through the NEPA process.

Provided this alternative is of comparable function and character to the destroyed facility, it is eligible for ER funding. This outcome rarely arises under the ER program. In almost all cases, it is practical or feasible to replace a damaged facility in-kind on its existing location. Casual or offhand application of the practicability and feasibility criteria to support any determination to replace on a new location is not acceptable; rather they must be applied as very rigorous tests.

For the first and third replacement scenarios discussed above, it is possible that the replacement facility proposed by the State could also be viewed as an added protective feature betterment eligible under the ER program. Replacement on a new location to reduce the potential for future disaster damage to the facility is an example. Of course, to be eligible for ER funding, the additional cost of the relocated facility must satisfy the economic justification criteria applied to added protective features. If the relocated replacement facility meets the economic justification criteria, then ER funding is not capped as discussed above but may participate in the cost of the replacement facility. This is illustrated by the following example:

Replacement at New Location Where Repair or Replacement In-Kind is Possible (ER Funding of Betterment)

A three-mile section of a non-Interstate highway in a low-lying coastal area was heavily damaged by high surf and wave action from a storm, the third time in the last 15 years that the highway has been damaged. The latest damage would cost $2 million to repair. Instead of repairing this section of highway, the State proposed to replace it on a new location further inland at a cost of $5.5 million, substantially reducing the opportunity for future surf and wave damage. Relocation of the facility is viewed as an added protective feature that needs to be economically justified in terms of the investment of ER funds. For this example, considering the frequency that events have occurred at this location, the added feature (relocation) was considered economically justified under the ER program, and ER funding may participate in 80 percent of the incurred costs of the replacement facility, or $4.4 million.

III. Emergency Relief Application Process

The decision to seek ER financial assistance rests with the State. Local highway agencies do not deal with the FHWA directly but must make their application through the State. The State has the option to determine whether it will seek ER funding for repair of either State or local agency owned Federal-aid highways.

The State and local transportation agencies are empowered to begin emergency repairs immediately to restore essential traffic service and to prevent further damage to Federal-aid highway facilities. Properly documented costs will later be reimbursed once the State formally requests ER funding, and the FHWA Division Administrator makes a finding that the disaster is eligible for ER funds.

There are two methods for developing and processing a State request for ER funding as described in this chapter. The first method is labeled as "Traditional" since it is the normal process used to develop a funding request. The second method, the "Quick Release" method, employs a process to immediately deliver ER assistance for large disasters very quickly. The quick release method should not be used as a matter of routine and is intended to provide a "down payment" on overall ER needs immediately following a large scale disaster.

A. Method 1 - Traditional

1) *Preliminary Steps* - The Governor issues a formal proclamation of the existence of a disaster. A Presidential declaration, or the Governor's request for this declaration, can serve the same purpose. The State then files a letter of intent to apply for ER funding

with the FHWA Division Office in the State, and the Division Office acknowledges the letter. These actions are described in Chapter IV.

2) *Disaster Assessment* - Detailed damage inspections are conducted at many, if not all, sites. Alternatively, initial damage assessments could be based on windshield surveys at a sampling of sites as described in Chapter V. When conducting windshield surveys, at least one site is visited in *each county* involved in the event. Windshield surveys may be used under this method to expedite the funding request process. If windshield surveys are used initially, detailed damage inspections should be completed at a later time. Since windshield surveys may not provide the required accuracy of damage funding needs, requests based on windshield surveys should be conservative with an additional follow-up request submitted based on more refined assessments. Disaster assessments are described further in Chapter V.
3) *Formal State Request for ER Funding and Damage Survey Summary Report* - The State requests ER funding based on the detailed damage inspections, the windshield surveys, or a combination of both. The State's request, along with the information from the site inspections and the FHWA Division Office's recommendations, are submitted for a finding by the FHWA Division Administrator. This process is described in Chapter IV.
4) *Length of Time to Develop Application* - Typically, the above steps would be accomplished over a 6 to 10-week period. If windshield surveys alone are used to expedite the process, then the above steps may be accomplished over a 2-3 week period.
5) *Division Administrator's Finding* - For events approved for ER funding, the Division Administrator issues a letter to the head of the State transportation agency notifying the State of the approval. Concurrently, the FHWA Division requests the FHWA Headquarters Office of Program Administration, either by memorandum or e-mail, to allocate ER funds. If the request is made by e-mail, it must be sent to the Office of Program Administration official mailbox (FHWA, ProgramAdministration) with a copy to the ER Program Manager. The request, as a minimum, must include date of approval and brief description of the event, estimated cost of damage repair and the Federal share, the ER obligation need for the current FY, list of counties, and the US Congressional district numbers (see sample request memorandum in Appendix A). Additional ER funding will be made available to the State as needs arise. The Division Administrator will notify the State transportation agency of any events that are denied ER funding.

B. Method 2 - Quick Release

1) *Preliminary Steps* - Same as Method 1, although the State's letter of intent and request for an ER finding may be combined into a single document.
2) *Disaster Assessment* - For the purposes of an ER application, few, if any, on-site damage surveys are made. Instead, the disaster assessment is based on other readily available information, such as credible media reports or aerial surveys done by the State.

3) *Formal State Request for ER Funding and Damage Survey Summary Report* - For the purposes of the ER application, the State requests ER funding based on their preliminary assessment of the extent of damage. The request is done in a brief letter to the Division Office. No Damage Survey Summary Report is prepared to accompany the ER application. A sample State letter is provided in Appendix F.

   The Division Office makes a finding of eligibility of the event based on the readily available reliable information on the extent and severity of the damage. The Office of Program Administration in Headquarters is available for consultation as needed. A request for an initial allocation of ER funds is made to the Office Of Program Administration, usually for a modest amount only intended to start the flow of ER funds to the State for initial emergency repair costs the State is incurring. Supplemental allocations would follow at later dates. The determination of an appropriate initial allocation amount is typically discussed by phone between FHWA Division Office staff and staff from FHWA Headquarters who administer the ER program.

   The State transportation agency should prepare and submit an abbreviated Damage Survey Summary Report to the Division Office after most of the detailed damage inspections have been completed. The report can be as simple as a summary of eligible estimated ER costs by county. This report must provide support for supplemental allocations of ER funds for permanent repairs as well as providing a summary of the counties involved in the disaster and an overall estimate of ER eligible repair costs.
4) *Length of Time to Develop Application* - Typically, the above steps would be accomplished over a 1 to 2-day period.
5) *Division Administrator's Finding* - Same as Method 1.
6) *Detailed damage Inspections* – Such inspections are completed at a later date, most likely after the Division Administrator has made a finding of ER eligibility for the event. These detailed damage inspections of sites at a later date aid in eligibility determinations and approval of a program of projects. At this point in time, it is likely that a refined estimate of ER funding need would be provided to FHWA Headquarters to facilitate a more accurate allocation of funds.
7) *Additional Considerations* - The "quick release" method is not appropriate for all disasters. In general, it is used for larger disasters where extensive damage is readily evident and where there is a desire, both by the Administration and a State, to have ER resources flowing quickly to the State.

C. Two Disasters Treated as One

Although unusual, disasters can occur so closely in time that the damage to roads and bridges is difficult or impossible to evaluate and tabulate as separate events. Many facilities in the same area may suffer damage from a second disaster before the survey teams have evaluated the damage caused by the earlier event.In these situations, the data supporting the magnitude and extensiveness of the second occurrence should be supported in the regular manner; but the request submitted should ask that the two disasters be recognized as a single event.

D. ER Program Flow Chart

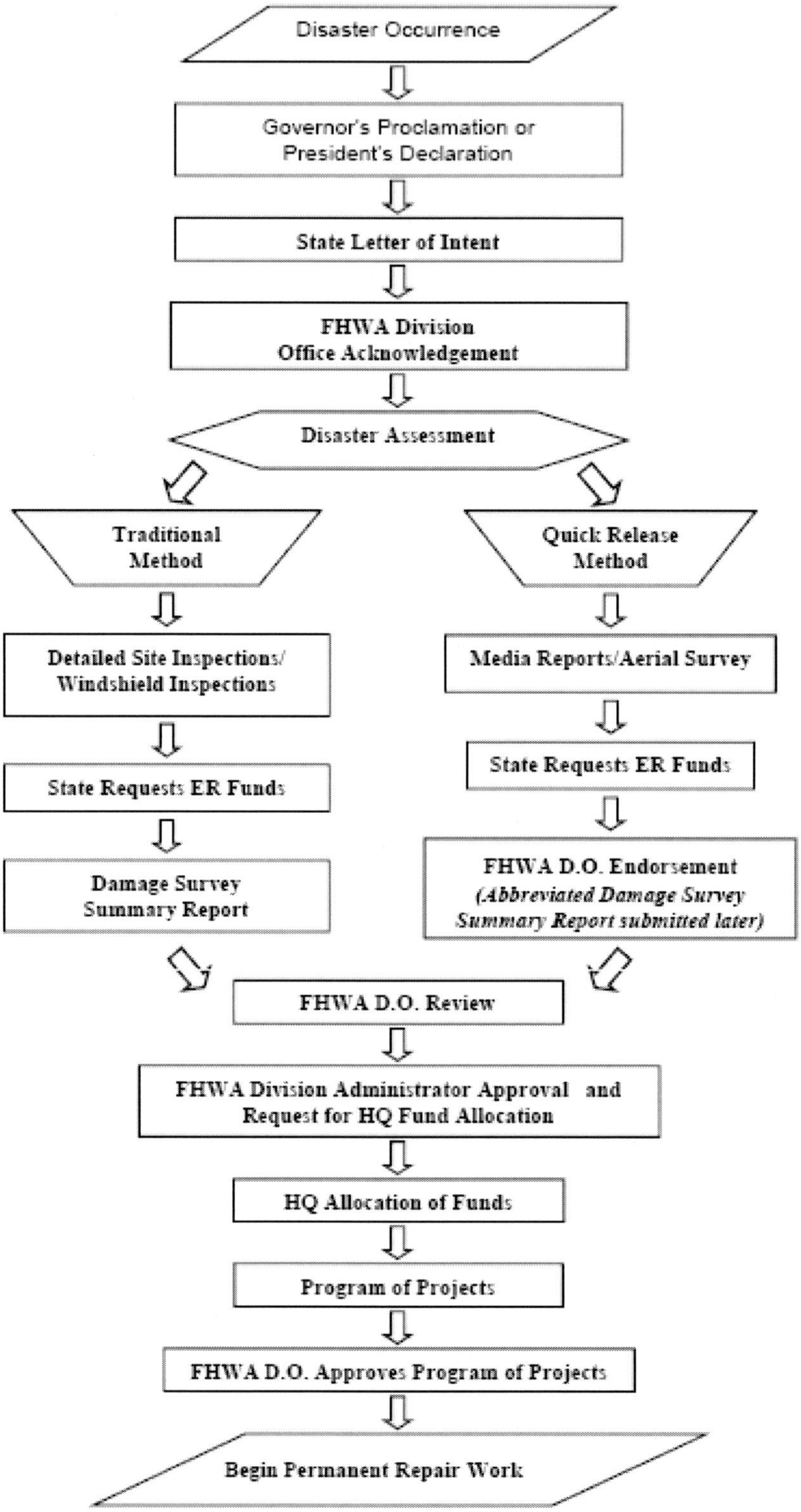

This may, of course, not be desirable if one or both disasters are potentially subject to the $100 million per State cap. In this instance, survey teams should make a "best estimate" of damage to be attributed to each event.

Also, it should be noted that where two disasters are treated as one, there is only one 180-day period covering the 100 percent Federal share for the combined event; this period beginning with the occurrence date of the first disaster.

Granting a request to treat two disasters as one will simplify the processing of the necessary repair or reconstruction projects by having a single ER finding for all projects resulting from the two disasters.

The Governor's second proclamation should cover both disasters. However, the supporting data must prove that facilities have suffered severe damage in both disasters and such damage is of sufficient severity to justify ER assistance. Sufficient documentation establishing the existence of separate disasters should assure compliance with the $100 million per State cap per disaster.

IV. Preliminary Steps

A. Letter of Intent

The State files a "letter of intent" with the FHWA Division Office notifying of its intention to request ER funds. This is usually done as soon as it is evident that there is eligible damage, either during or immediately after the occurrence of the disaster. Filing a letter of intent does not commit the State to any future course of action.

The letter of intent is simply a notice that a State plans to seek ER funds and does not need to specify the amount or other details. An example of a typical letter of intent is contained in Appendix B. An electronic copy of the letter of intent should be sent to the Office of Program Administration with a copy to the ER Program Manager.

B. Acknowledgment Letter

The FHWA Division Office acknowledges in writing the State's letter of intent. An example of a typical Division Administrator's reply to the State's letter of intent is contained in Appendix C.

The acknowledgment letter to the State should address the following items: acknowledgment of receipt, effective date, emergency operations, preliminary engineering, use of force account, waiver of competitive bidding, guidance for permanent work, time frame and items needed for Damage Survey Summary Report, time limit for programming, payment contingent on approval by the Division Administrator, and record keeping requirements.

The Division Administrator's acknowledgment of the applicant's letter of intent will assure the State that temporary operations, emergency repairs, and preliminary engineering may proceed without prior authorization by the FHWA. However, the eligibility of such work for ER funds will be contingent upon a favorable finding by the Division Administrator on the eligibility of the disaster and subsequent approval/authorization of the work by the FHWA. An electronic copy of the acknowledgement letter should be sent tothe Office of Program Administration with a copy to the ER Program Manager.

C. Governor's Proclamation

This is a formal proclamation by the Governor of the existence of an emergency. An example of a Governor's proclamation is contained in Appendix D. Only those areas of the State that are included in the Governor's proclamation will be eligible for ER funding.

1. Basic Criteria

When serious damage has occurred or is being caused to Federal-aid highways, the Governor of the State shall issue an official proclamation unless the Governor has requested a major disaster declaration by the President under the Disaster Relief Act (P.L. 93-288). The proclamation should be issued by the Governor during or shortly after the occurrence and should state the gravity of the situation and specify the area affected. In the case of a catastrophic failure, it must be established that a catastrophic failure has occurred, (e.g., a sudden failure of a major element or segment of the highway system that causes a disastrous impact on transportation services) and that the cause was external to the facility.

2. Timing of the Proclamation

Although the Governor's proclamation may be issued during or with reasonable promptness following the event, there is no need from FHWA's standpoint for the proclamation to be issued until a realistic determination has been made as to the area affected. It is desirable to delay preparation of the Governor's proclamation to permit evaluation of the magnitude and extent of the natural disaster, or to evaluate the magnitude of the catastrophic failure, by engineers from the State and FHWA. This procedure should facilitate consistency between the list of counties in the proclamation and those covered by the Damage Survey Summary Report. A copy of the Governor's proclamation shall be obtained for the Division Office file.

3. Relationship to President's Declaration

The proclamation by the Governor as required in Title 23 is an entirely separate official action from the declaration issued by the President of the United States. The President's declaration proceeds from the disaster relief authority of P.L. 93-288 and is in response to a request from the Governor. The Governor's request for a determination by the President may be submitted as part of the State's application for ER funds in lieu of a State proclamation. A copy of the Governor's request shall be obtained for the Division Office file.

4. Concurrence by the Division Administrator

The FHWA's concurrence in the disaster eligibility is either based on the emergency proclamation by the Governor, or, when the Governor has requested a major disaster declaration by the President and the declaration has not been made, on a review of the State's submission required for the ER application as described in Chapter III. The Division

Administrator's concurrence will indicate his/her finding that extraordinary natural disaster conditions did exist, or that a catastrophic failure occurred, and that the disaster is eligible for ER funding assistance under 23 USC. 125.

When the President has issued a major disaster declaration, the Division Administrator need not concur that a disaster occurred, but must nevertheless make a finding, based on a review of the State's submission required for the ER application in Chapter III, that substantial damage has occurred to the Federal-aid highways over a wide area or that the criteria for a catastrophic failure were met and that the damage is eligible under 23 USC. 125.

V. Disaster Assessment and Damage Survey Summary Report

A. Purpose

For a natural disaster, the disaster assessment is an evaluation of a natural occurrence that affects transportation facilities to determine if the basic conditions exist to support a reasonable request for ER funding. In any natural disaster situation two conditions must be met in order to qualify for ER funding: (1) the natural occurrence is sudden, unusual, and causes serious damage to Federal-aid highways; and (2) the extent of serious damage to Federal-aid highways covers a wide area. The natural occurrence over a wide area must have inflicted unusual heavy economic loss to the State and its subdivisions or other organizations or agencies.

Disaster assessments for catastrophic failures must provide an evaluation of the failure against the basic requirements for ER funding eligibility. The sudden failure of a major element or segment of a Federal-aid highway due to an external cause is defined as a catastrophic failure, and the disaster assessment must gather, organize and discuss the necessary data to either support or rule out a catastrophic failure event. The failure must not be primarily attributable to gradual and progressive deterioration, or lack of proper maintenance. The closure of a facility because of imminent danger of collapse is not considered a sudden failure.

B. Logistics/Mobilization

Experience dictates the need for developing a system for documenting damage estimates related to a disaster event and for the preparation of the Division Office Field Report. A good procedural method consists of the following elements:

1. Disaster Coordination Engineer

One engineer from the Division Administrator's staff should be selected and delegated the responsibility for coordinating the Division Office activities relating to disaster assistance programs. This approach should provide uniformity in administration of all ER program activities.

2. Division Office Orientation

The Division Disaster Coordination Engineer should arrange a training meeting to brief field engineers on resource and procedural requirements necessary to fulfill Division Office responsibilities. Eligibility criteria and field reporting procedures should be discussed and review/evaluation teams organized. The following items should be provided to each FHWA engineer charged with assessment responsibility:

- Transportation agency contact's name
- Emergency Relief Manual
- Camera and film
- Maps of affected areas showing Federal-aid routes
- Calculator and Measuring Tape
- List of unit prices
- Laptop Computer
- Communication equipment including Cellular Phone
- Detailed damage inspection report forms (Appendix E)

3. Resource Evaluation

An initial evaluation of staffing/personnel requirements, equipment, and financial needs should be made. These resources must be adequate for the timely completion of the required disaster assessments and damage surveys. For example, concerning personnel needs, each FHWA Division Office should confirm that driver's licenses are current and assess possible immunization requirements.

The disaster assessment should be completed quickly to permit submission of the required information for the ER request within the prescribed time-period. Combining the detailed inspections with the disaster assessment may take longer and cause a late assessment.

C. Coordination with other Agencies

Most natural disasters involve several Federal, State, and local agencies. It is necessary and often critical to establish immediate contact with these agencies to expedite the ER program. Coordination with the Office of Federal Lands Highway Programs is of particular importance since they maintain close liaison with the US Forest Service, Park Service, Bureau of Indian Affairs, and the Bureau of Land Management regarding damage to Federal roads that may have sustained damage from the same event. Also, coordination with appropriate environmental resource agencies is encouraged. Among other reasons, coordination can avoid embarrassing inconsistencies and inaccurate total damage estimates.

Meetings set up by the State transportation agency are often advantageous, particularly if they are similar to, or in conjunction with, those held by the FEMA. Since the State makes the request for ER funds, it has the responsibility to arrange these meetings. The FHWA should provide technical assistance at these meetings, as requested.

These meetings should provide briefings of all eligible applicants and personnel who will be involved in ER projects. Participants should include State maintenance forces, key State transportation headquarters and division and/or district office personnel, FHWA field engineers, administrative staff from local governments, and representatives from Federal agencies. Eligibility, field procedures, and Damage Survey Summary reporting procedures should be discussed.

Applicants should be advised of the necessary records and documents that must be available to support expenditures on ER projects. Permanent restoration work must follow normal Federal-aid procedures.

1. Policy

The Federal Government will participate in costs incurred by a State transportation agency, or a political subdivision of that State, when such costs are properly supported and are directly attributable and properly allocable to ER projects. Needed documentation, as applied to the several categories of cost, is discussed later in this chapter.

2. State Transportation Agency

All requests for reimbursement shall be submitted consistent with procedures followed in billing other types of Federal-aid project costs. Billings for reimbursement will be subject to audit by State and Federal representatives. (Reference *Federal-Aid Policy Guide*, 23 CFR Part 140). A billing for eligible costs incurred by the State should be submitted to the FHWA for reimbursement. A billing for participating costs incurred by a political subdivision of a State should be submitted to the State which, in turn, may submit a billing for such costs to the FHWA for reimbursement. Final billing should be submitted promptly after the final inspection of the completed work. Progress billings are reimbursed on permanent repair projects. All billings should result from the project cost records and the accounting system. In order to obtain maximum reimbursement from the Federal Government, any political subdivision expecting to incur ER project costs should obtain guidance in the preparation and maintenance of supporting documentation as well as in billing procedures from officials of the State. Source documents are to be preserved for a period of at least 3 years after payment of the final voucher by the FHWA.

D. Damage Assessments

The State, in cooperation with FHWA field staff and in coordination with local authorities, undertakes damage assessments. Typically, this involves on-the-ground visits to the damage sites to verify the extent of damage and to estimate the cost of repairs eligible for ER funding. If there is a need to expedite the delivery of ER funding for high-cost disasters, an initial damage assessment may be based on aerial fly-overs, news telecasts, and other means of quickly estimating the extent of damage. This initial damage assessment is followed-up later with more detailed site inspections.

As discussed in Chapter III, the damage assessments are conducted using one of two different methods, with the choice of method depending upon the urgency for developing and processing a State's request for ER funding for a disaster or catastrophic failure eligible under the FHWA ER program. Under the Traditional Method, detailed damage inspections are conducted to document site-by-site estimates. Alternatively, expedited windshield inspections may be conducted to document estimates for a limited number of sites. Under this method, a combination of detailed inspections and windshield inspections may also be employed. Under the Quick Release method estimates are based on readily available information such as valid media reports or aerial surveys done by the State.

1. Detailed Damage Inspections

These are inspections conducted on site to determine the extent of damage, scope of repair work, the preliminary estimate of cost of repair, and ER funding eligibility. Detailed damage inspections are conducted at many, if not all, sites. After it becomes apparent that the Division Administrator will recommend a positive determination of natural disaster, disaster inspection teams should be organized.

These teams normally document site-by-site repair estimates to develop supporting material for programming purposes. This detailed inspection may be accomplished in conjunction with the disaster assessment if it does not delay submission of the Damage Survey Summary Report.

The on-site inspection provides an important opportunity to define clearly the extent of repairs eligible for ER funding. Eligibility determinations under the ER program reside with FHWA.

a. Documentation

The damage inspection report should document:

- The specific location, type of Federal-aid highway, ADT, cause, nature and extent of damage, including mileposts where available
- The most feasible and practical method of repair, particularly if the applicant's proposal does not meet eligibility criteria (See Chapter II)
- Work considered to be emergency or permanent should be identified and documented
- The estimated repair cost
- Recommendation by the FHWA field engineer
- Acknowledgment by the applicant (and State representative, as appropriate)
- Potential environmental/historical impacts
- Photographs supporting the above
- A location map and field site sketch
- A blank detailed damage inspection report is provided in Appendix E.

b. Disaster Inspection Teams

The inspection teams consist of representatives from the FHWA and the applicant. If the applicant is not the State transportation agency, a State representative should be a member of the inspection team. Specialists (for example, bridge engineers if there is significant bridge damage) from each agency should also accompany the team if the situation warrants. Other specialists in such areas as right-of-way, environmental evaluation, or geo-technical analysis may also be needed, depending on site conditions.

The applicant's representative identifies the cause of damage and the normal design and construction practice to repair the facility. The FHWA field engineer shall consider site eligibility and proposed repair effort to aid in developing an estimate. All sites reviewed by the applicant should be documented with the FHWA eligibility recommendations noted. If all or any of the parties cannot agree on eligibility, the FHWA representative must note all items of disagreement on the inspection report. The applicant may appeal an ineligibility finding that should be submitted in writing to the Division Administrator within 30 days after the initial finding. The FHWA representative should also explain that any ER eligibility or participation disagreements are decided by the FHWA Division Administrator, or, when necessary and appropriate, the FHWA Administrator.

Copies of the inspection report should be provided to the applicant and members of the inspection team.

As a general practice, FHWA should participate in the damage inspections. However, events with a large number of damage sites and multiple inspection teams in the field at one time may overextend FHWA staff resources.

If an FHWA representative cannot be present on each inspection, the FHWA inspection workload should be ordered by such priority considerations as project cost, scope of work, good agency record keeping and documentation, the jurisdiction whose sites are involved, and the general experience and capabilities of the inspection teams. In addition, other steps that occur during administration of an ER disaster, such as the preparation of a program of projects or final inspections that may be undertaken by the Division Office, afford the FHWA some opportunity to further assess eligibility issues.

In view of the ER eligibility issues involved, decisions concerning any site inspections that will not be attended by FHWA must be based on appropriate risk management principles.

2. Windshield Inspections

Where appropriate, windshield surveys are used to expedite the ER application process. These reviews verify the extent and impact of damage during or immediately following a disaster and collect damage information to determine disaster eligibility for ER funding. Damage assessments are based on a windshield inspection at a sampling of sites as described below, but at least one eligible site must be visited in each county involved in the event.

The State transportation agency and Federal or State agencies with responsibility for highways under their jurisdiction will have advance information on the initial damage, road closures, and in some cases a very rough estimate of costs to restore facilities. As part of the more general disaster assessment, FHWA field engineers will be assigned to verify extent and severity of damage to highways and bridges.

In certain circumstances where time permits and the number of sites can be adequately sampled, ER assessments may coincide with assessments required by FEMA. Concurrence by FEMA should be obtained in advance.

a. Scope of Review

An independent verification by FHWA is required. Depending on the time available, amount of damage, how widespread the damage is, and accessibility, FHWA field engineers must review as much of the identified or reported damage as possible. As noted above, at least one eligible site in each county recommended for ER funding should be verified. This verification will assure the Division Administrator that no jurisdiction will be promised funds where not eligible and that the most critical needs are approved at the earliest practical time.

For an obvious situation, such as a massive bridge collapse, these conditions may be immediately met. For a marginal condition, many weeks of verification may be necessary.

The Division Disaster Coordination Engineer should be responsible for prescribing to the FHWA field engineers the format and extent of information necessary to prepare the Damage Survey Summary Report and to support the Division's recommendations.

b. Record-keeping

As a minimum, the Division's file should contain information on the extent of and methods used to evaluate the disaster and copies of the FHWA field engineer's assessments on damage and estimates of cost.

c. Supplemental Information

Types and sources of other information and data that should be considered if available:

- Damage reports from other agencies supported by photographs or field verification by FHWA personnel
- Newspaper articles
- Photos, including aerial photos
- US Weather Bureau data
- US Geological Survey information
- Reports from others, e.g., flood control agencies, cities, public utilities, Corps of Engineers, etc.
- Interviews with local citizens
- Contact with FEMA where a Presidential declaration is involved.

Detailed damage inspections of sites are completed at a later date, likely after the Division Administrator has made a finding of eligibility for the event. These detailed damage inspections of sites at a later date aid in eligibility determinations and approval of a program of projects.

3. Damage Assessments for Quick Release

Quick Release damage assessments are based on readily available information such as valid media reports or aerial surveys done by the State. Detailed damage site inspections are conducted at a later date, most likely after the Division Administrator has made a finding of eligibility for the event.

E. Damage Survey Summary Report

The FHWA Division Office assists the State transportation agency in preparing a Damage Survey Summary Report based on the State's assessments. The Damage Survey Summary Report provides the Division Administrator with a basis to make a finding that the disaster is eligible for funding under the FHWA ER program.

1. Purpose

The purpose of the Damage Survey Summary Report is to summarize the damage assessment and provide information and documentation for the FHWA Division Administrator to make a finding that a natural disaster or catastrophic failure has occurred within the intent of 23 USC. 125. The report describes the general nature and extent of the resulting emergency situation and delineates the limits of serious damage to Federal-aid highway facilities. The sudden and unusual nature of the disaster should be documented and evidence of external cause should be included for a catastrophic failure.

A long-term problem, e.g., a very slow moving slide or subsidence or a slow lake rise, is considered outside the scope of a "disaster", since necessary work is more accurately categorized as preventive rather than repair. Serious damage for purposes of supporting a finding of ER eligibility is heavy, major, or unusual damage to the highway that severely impairs the safety or usefulness of the highway or results in road closure. Serious damage requires more than normal heavy maintenance to repair. Examples of serious damage include destroyed bridges, damaged bridges incapable of supporting traffic, loss of traffic control devices causing severe disruption, or major slides and slip-outs extending into the traveled way. Applications for ER funds in amounts less than $700,000 must include a statement explaining why the damage repair involved is beyond the scope of heavy maintenance or routine emergency repair. Generally, widespread nominal road damages in this range are not considered significant and therefore do not justify approval by the FHWA Division Administrator for ER funding.

2. Damage Survey Summary Report Preparation/Content

The Damage Survey Summary Report should include:

a) A description of the type and extent of damages and the estimated cost of restoration or reconstruction by Federal-aid routes for each county. In addition, the Division should provide the Federal share of the estimated cost of repair work and determine the amount of ER funds needed for repairs during the current FY.
b) A description of the limits of the areas involved and the nature and characteristics of the disaster or catastrophe including the dates of occurrence. This information will differentiate between ordinary and extraordinary natural disturbances, except when the President has declared that a major disaster exists over the area involved.
c) Photos showing the extent of serious damages sustained in the areas being recommended. At least one photo showing eligible damage should be included for each affected county.

3. Damage Survey Summary Report Submission

The Damage Survey Summary Report and supporting documents shall be submitted to FHWA Division Office.

4. Summary of Other Required Documents

If not previously submitted, the following documents must accompany the Damage Survey Summary Report:

- Governor's Proclamation, or a copy of the Governor's official request for a Presidential disaster declaration
- A copy of the State's written request for ER funds

5. Exception

Formal preparation of a Damage Survey Summary Report is not required when requesting ER funds using Method 2 - Quick Release, as described in Chapter III. For the purposes of the ER application, the State requests ER funding based on their preliminary assessment of the extent of damage via a brief letter to the FHWA Division Office. No Damage Survey Summary Report is prepared to accompany the ER application. The Division Administrator will then make a determination of eligibility of the disaster for ER funding. However, the Division Office should coordinate with the FHWA Headquarters ER Program Manager prior to the eligibility determination. Refer to the topic Quick Release under Chapter III. An abbreviated Damage Survey Summary Report should be prepared and submitted to the FHWA Division Office upon completion of a sufficient number of detailed damage inspections to demonstrate that an eligible disaster event has occurred. A Damage Survey Summary Report is still needed to meet the requirements of preparing a program of projects for final approval by the Division Administrator.

VI. Project Procedures and Requirements

A. General

Once the FHWA Division Administrator has made a finding that emergency or catastrophic conditions justify ER funding, the State should submit promptly a program of projects for repair of damage to the Federal-aid highways. Projects should be individually justified. If sufficient information is available when the Damage Survey Summary Report is submitted, the first program of ER projects may be incorporated into the Damage Survey Summary Report itself, or it may accompany that report, even though a finding has not yet been made. In any case, the program of projects should be submitted within three months after the disaster finding by the FHWA Division Administrator.

A program of projects is to be submitted to the FHWA Division Office regardless of the Division Office's role in project oversight on Federal-aid projects.

B. Fund Management

FHWA's Office of Program Administration will maintain a current table of nationwide ER requests. This table will be updated as requests for ER funds are received. This table will consist of "formal requests" and "pending requests" (described below) based on the stage of the application process. A "pending request" should be developed shortly after an ER-eligible event has occurred. The FHWA Division Office should work with the State to develop an initial "best estimate" of ER needs for the event. Such an initial request should be provided to the Office of Program Administration as soon as practicable for inclusion on the current ER request table. Pending requests should be updated as better estimates are developed. Such pending requests are subject to change and will not result in an actual allocation of ER funds. However, it is important that the Division Offices keep the Office of Program Administration apprised of the most current needs for any known ER eligible events. At times, Congress may choose to provide a supplemental ER appropriation if current available funding falls short of total outstanding ER needs. Maintaining the most current ER request data allows Congress to accurately address all outstanding ER needs if a supplemental appropriation is considered.

After the FHWA Division Administrator makes an affirmative finding on a State's request for ER, the Division requests an allocation of ER funds from the FHWA Office of Program Administration (see Chapter III, *Emergency Relief Application Process*). This request represents the "formal request" for an actual allocation and will be included on the current ER request table. This allocation request is based on the State's anticipated ER obligations for the current FY and may be less than the total ER needs for that event. The actual allocation amount may be less than the amount requested, depending on the availability of ER funds. FHWA's Office of Program Administration will allocate additional ER funds based on outstanding nationwide ER needs, provided additional funds are available.

The FHWA Division Office should work closely with their State to accurately determine ER obligation needs for the current fiscal year. It is recognized that funding requests may occur at any time during the FY. A funding request near the beginning of the FY will allow for better estimating of current FY obligation needs. However, for funding requests occurring late in the FY, it may be more difficult to estimate obligation needs for the remainder of the FY. Divisions are encouraged to work with the State to prepare an annual finance plan, as appropriate, that will facilitate an accurate estimate of ER obligation needs during the FY.

In addition to pending and formal requests submitted as described above, the Office of Program Administration will request from all Division Offices a list of ER needs (both pending and formal). This request will usually be made near the end of each FY, but may be made at other times as well. The identified needs will be added to the nationwide ER request table in preparation for an allocation of funds. Near the end of each FY, the Office of Program Administration will also coordinate with Division Offices to identify any balances of previously allocated ER funds that will not be obligated through the remainder of the FY. Those funds will be withdrawn to be used for other nationwide ER funding needs during the next allocation. This process is intended to avoid accumulating large balances of unobligated ER funds and helps manage available funds nationwide as effectively as possible.

C. Federal Share

The Federal share for the repair of Federal-aid highways is established by law. It varies depending on the nature of repairs, when the work is accomplished, and the Federal-aid route being repaired.

For the costs associated with restoring essential traffic, minimizing the extent of damage, or protecting the remaining facility which are incurred in the first 180 days after the occurrence of the disaster, the Federal share is 100 percent. For the costs of permanent restoration work, and the costs of all repairs incurred after the first 180 days, the Federal share is based on the type of Federal-aid highway that is being repaired. For Interstate highways, the Federal share is 90 percent. For all other Federal-aid highways, the Federal share is 80 percent. The Federal share can be increased based on the "Sliding Scale" rates in States with high percentage of Federally owned public lands. The Federal share is 100 percent for all work done on roads on Federal lands.

D. Preparation and Submission of Programs

A program of ER projects should be prepared by the State. The program of projects should:

- Indicate the natural disaster or catastrophic failure and the time of its occurrence.
- Relate the damage to that described in the damage assessment reports prepared and/or detailed damage inspections.
- Describe proposed permanent repairs or replacements on a site-by-site basis (although sites may be lumped by route and county for program purposes).
- Include supporting material indicating the suitability and economy of upgrades or betterments including relocation proposed for participation with ER funds. For some projects it will be necessary to complete additional design work in order to develop justification for added protective features. When betterments are contemplated, the State or local agency should contact the Division Administrator so that further project development is accomplished with FHWA involvement.
- Identify emergency repairs.

E. Approval of Programs and Project Authorizations

The Division Administrators have authority to approve programs of projects that are located on Federal-aid highways.

Temporary operations including emergency repairs, and preliminary engineering, including consultant work, may proceed without prior authorization.

This work need not be authorized retroactively; however, the need for such work must subsequently be approved by the FHWA as part of a program of projects. Permanent restoration work shall not be performed prior to FHWA authorization unless performed as part of emergency repairs.

F. Advancing Projects during ER Program Funding Shortages

When ER funds are not available for allocation to the States to cover either additional funding needs on previously approved ER events or funding needs for new disaster requests awaiting action by the Division Administrator, ER funding requests received in Headquarters are recorded and held by the Office of Program Administration pending action by Congress to replenish the ER accounts through a supplemental appropriation.

Recognizing that quick congressional action is not always possible, the following options could be used to fund or advance ER projects on an interim basis.

1. Previously Approved ER Events

For an event that the Division Administrator has previously found eligible for ER funding, requests to fund additional work for that event could be handled under one, or some combination, of the following two options:

a. Option 1 - Use of Regular Federal-Aid Highway Funds

Regular Federal-aid highway funds, appropriate for the type of Federal-aid highway (National Highway System or Surface Transportation Program), can be used. Regular Federal-aid funds must comply with the obligation limitation in effect for the class of funds used. The Federal share would be that appropriate for the ER work being authorized. Under this option, the letter of authorization should indicate that the project will be converted to ER funding when ER funding becomes available, at which time the regular Federal-aid funding, and the accompanying obligation limitation, will be released from the project.

This option can be used for both emergency repairs to restore essential traffic as well as permanent repairs.

This option has the advantage of allowing immediate Federal reimbursement for costs that are being incurred. Further, it provides a means of securing FHWA authorization of the permanent repair activities so that they may proceed. A disadvantage is that use of regular Federal-aid funding sources will likely require the use of obligation limitation until the project can be converted to ER funding.

b. Option 2 - Use of Advance Construction (AC)

Although 23 USC. 115 does not contain authority to advance construct ER funds, it does designate several other Federal-aid funding sources that can be used for advance construction.

Thus, a project advance constructed under any of the funding sources designated in Section 115 could later be converted using ER funds.

For example, an ER type project can be authorized as an advance construction STP project and later converted to an ER project as ER funds become available. An authorization under this option should be made following the general guidance for advance construction of Federal-aid. The letter of authorization should confirm the State's intention to convert the project to ER funds.

This option can be used in those instances needing prior FHWA authorization of permanent repair work. This option has the advantage of providing a means of securing FHWA authorization of the permanent repair activities so that they may proceed.

Further, it does not use obligation limitation. However, a State must have adequate funding resources of its own to proceed with the project until it can be converted to ER funding.

2. ER Events Awaiting a Division Administrator Finding

When a potential ER event has occurred, the State is empowered to undertake immediate emergency repairs to restore essential traffic service and to prevent further damage to Federal-aid highway facilities. Properly documented costs will later be reimbursed once the Division Administrator makes a formal finding that the event qualifies for funding under the ER program.

However, if a delay in the Division Administrator's formal finding will likely delay the orderly progression of both emergency and permanent repairs, it might be preferable to proceed with ER activities using Options 1 or 2 above, subject to the following requirements:

- The State's formal request for ER funding, along with an acceptable Damage Survey Summary Report has been submitted to the FHWA Division Office. Various methods for developing State requests are discussed in Chapter III.
- Division staff managing the ER program has reviewed the ER request and recommends that it be approved, excepting any questionable or ineligible ER activities.
- FHWA letter of authorization under Options 1 or 2 will stipulate that any use of ER funding on the project is subject to the Division Administrator's formal finding that the event qualifies for funding under the ER program.

Options 1 or 2 should be limited to specific events where the delay in securing a formal finding is lengthy and is delaying repair efforts. Any Division Office authorization of work prior to the Division Administrator's formal finding should be coordinated with the Office of Program Administration.

3. General Comments

Prior to FHWA authorization of permanent ER repairs using regular Federal-aid funds or the advance construction process, it is strongly recommended that the Division Office review the activities and project features to be funded to assure their eligibility before starting construction.

G. Project Oversight

ER projects for permanent repairs should be processed following regular Federal-aid procedures. ER projects the same as or sufficiently similar to regular Federal-aid projects subject to the 23 USC. 106 oversight exceptions can also be administered under these exceptions, subject to the following two conditions:

1. Any betterments to be incorporated into the project and for which ER funding is requested must receive prior FHWA approval.
2. The FHWA reserves the right to conduct final inspections on all ER projects. The Division Administrator has the discretion to undertake final inspections on ER projects as deemed appropriate.

H. Combined Federal-Aid and Emergency Relief

When the State or applicant decides not to replace a damaged facility in-kind and proposes work in excess of the work eligible for ER funds, a combined project may be programmed using ER funds to the extent eligible. Other Federal-aid funds may be used for the additional work. Separate programming is required for each class of funds with appropriate cross-referencing.

I. Construction Start Deadline (Time Extensions)

ER funds are allocated to assist the States and other agencies or organizations in conditions of emergency. Consequently, after approval of programs and allocation of funds, all projects should be completed promptly. Failure to advance an approved ER project to completion within a reasonable period of time could result in withholding of funding for that project. Emergency opening work should be accomplished within one month of accessibility to the site under normal circumstances. Unless there is satisfactory justification for project delay to warrant its retention, projects for permanent repairs that have not advanced to construction obligation by the end of the second FY following the year in which the disaster occurred cannot be authorized. Justification for such delay and request for time extension must be submitted to the FHWA Division Administrator for approval. Time extensions are granted in one-year increments. Such delays may be caused by the need for extensive environmental evaluation, litigation, or complex right-of-way acquisition. In certain situations the delay of permanent work may be as much as two to three years. Permanent restoration work, for example, could be deferred to permit study of a serious slide condition, thereby allowing sufficient time to adequately design a permanent correction.

J. FHWA as the Construction Agency

State or local agencies may request the FHWA to accomplish repairs, reconstruction, or relocation of sections that are on the Federal-aid highways. The emergency operations to restore essential traffic should be handled by the State or local agency. In any event, where such situations are anticipated, a letter of request should be prepared by the State or local agency through the State to the Division Administrator expressing the desire to have the FHWA perform the work. The Division Administrator should promptly forward any such request to the Federal joint field inspection by the two offices, the local agency, and/or the State.

K. Project Designations and Numbering

All ER-funded projects on a Federal-aid highway (not on a Federal road) shall be designated with the prefix "ER". ER projects located on Federal Roads use the prefix "ERFO" (Emergency Relief Federally Owned). Combination projects designated "ER-ERFO" may be used where portions of a project on a Federal-aid highway are also located on a Federal road.

The State may designate the project numbering system to be used for each project resulting from a natural disaster or catastrophic failure. Where an existing "ER" series has been established, the State may continue the sequence of the established series of project numbers for several individual improvements, with separate agreement numbers for individual improvements.

Projects may be numbered to conform to the system established for other Federal-aid projects. For the project number, enter seven digits (four digits for Route number and three digits for agreement number) preceded by the prefix ER.

L. Disaster Code

Division Offices should be prepared to readily identify obligations by appropriation and by disaster. The Fiscal Management Information System (FMIS) provides a disaster number entry, which should be carefully entered to ensure that legislated limits on obligations for a particular disaster are not exceeded.

The disaster number is assigned by the Division Office at the time of the finding. This number should be shown as part of the disaster identification on the program. It consists of the two-letter State or Territory code, the FY of the initial damage, and the sequence number (based on the number of disasters submitted by a State) of the disaster. Thus, the first disaster submitted by Alabama for FY 2009 (i.e., the event began on or after October 1, 2008), would be shown as "AL09-1," with any subsequent eligible disasters submitted during FY 2009 following in sequence as "AL09-2," etc.

Item number 42 in the FMIS should be completed as follows: The Disaster Code (five digits) should be used. The first four digits represent the FY assigned to the event, e.g., 2009 or 2010, etc. Enter as the fifth digit the sequence number assigned to the disaster. For example, the first disaster approved in a State for FY 2009 would be coded 20091.

M. Construction Contracts/Force Account

Emergency repairs are temporary traffic operations undertaken during or immediately following the disaster occurrence for the purpose of: 1) minimizing the extent of the damage, 2) protecting remaining facilities, or 3) restoring essential traffic. All other repair or restoration work is considered to be a permanent repair.

Under 23 CFR 668.105(i), emergency repair work may be accomplished by the contract, negotiated contract, or transportation agency force account method as determined by the transportation agency as best suited to protect the public health and safety. Permanent repair and reconstruction work must be done by a competitive bid contract method unless the State demonstrates some other method is cost effective as described in 23 CFR 635.204.

1. Emergency Repairs

States shall advertise the work for emergency repairs where feasible. The FHWA may approve a waiver of the advertising requirement if State or local law authorizes such procedures and the contract method chosen is suitable for the proposed corrective work. Where time and conditions warrant, States are strongly encouraged to first consider using the competitive bidding method of contracting for emergency repairs.

Often, emergency repair work, such as debris removal, can be performed efficiently through the deployment of pre-established emergency repair contracts. Such contracts should be competitively bid and must comply with all applicable Federal-aid contracting requirements at the time of the disaster. States are encouraged to work with their FHWA Division Office to develop and periodically update pre-established emergency repair contracts prior to the occurrence of a disaster.

For States that have pre-established emergency repair contracts, FHWA Division Offices should work with their respective States to ensure that these contracts are up-to-date.

a. Force Account

The term force account means the direct performance of highway construction work by a State transportation agency, a county, a railroad, or a public utility company by use of labor, equipment, materials, and supplies furnished by them and used under their direct control. Public agencies may perform emergency repairs under the force account method, but are not permitted to compete for solicited or negotiated contracts. Since the National Guard does not fall under any of the above-mentioned categories, ER funds cannot be used to pay for National Guard services on a force account basis.

Due to the emergency character of the work, State and local forces and/or negotiated equipment rental contracts may be used to perform a considerable portion of the emergency repairs. In accordance with 23 CFR 635.204(b), a formal finding for force account work for emergency repairs is not required.

b. Solicited Contract

A solicited contract may be warranted due to the emergency character of the work. The State may contact a reasonable minimum number of contractors by telephone to solicit quotes for a specific scope of work. A summary showing how the solicitation was conducted, who was contacted, and the responses by the contractors must be prepared.

c. Negotiated Contract

Under certain emergency circumstances where it is critical to restore essential travel in an expedited manner, it may be appropriate to enter into a negotiated contract with one firm. The contracting agency must document the process it used for selecting and negotiating a reasonable price with a single firm. States are encouraged to use negotiated contracts only when the State determines that the circumstances are such that competitive bidding is not effective or feasible. Lump sum contracting should be used only when unusual or rare circumstances are present making it virtually impossible to estimate quantities of work for a competitively bid unit-price contract or a cost reimbursable negotiated contract.

The Division Administrator shall determine whether the price of a negotiated contract is reasonable under the circumstances of the situation. Where feasible, the State should conduct a cost analysis prior to the award of any negotiated contract to assure that prices are fair and reasonable to aid in this determination. Pursuant to 23 USC. 106, which requires the Secretary's approval of all plans, specifications, and estimates, the FHWA reserves the right to withhold funding or to reduce its participation when prices are not deemed to be reasonable.

A State should adopt appropriate industry rate guides, such as the Rental Rate Blue Book for computing equipment usage rates for negotiated contracts, or develop its own guide. The State must make the determination that the equipment rental rates developed or adopted fairly estimate a contractor's actual cost to own and operate the equipment, and the Division Administrator must concur in that determination. In situations where the rate of a particular item of equipment is not provided in the adopted guide, the State must determine that the rate is a reasonable representation of the contractor's actual cost to own and operate the equipment in light of the particular circumstances. If feasible, the State may compare the rates charged for using the piece of equipment in other projects for other entities, but taking into account the particular circumstances of the situation in which the State needs to use the piece of equipment.

2. Permanent Repairs

Permanent repair and reconstruction work, not accomplished as emergency repairs, must be done by contract awarded by competitive bidding unless the State demonstrates some other method is cost effective as described in 23 CFR 635.204. The contracting agency must assure an opportunity for free, open, and competitive bidding, including adequate publicity of the advertisements or call for bids. However, in certain cases, ER construction projects can be accelerated using other contracting techniques described below.

3. Techniques to Accelerate Projects

Innovative contracting procedures available to accelerate ER construction projects include cost-plus-time bidding, lane rental, construction manager at risk and design-build contracting. Other methods such as abbreviated plans, shortened advertisement period for bids, and incentive/disincentive clauses are commonly used to accelerate ER construction projects.

a. Cost-Plus-Time Bidding

Cost-plus-time bidding, more commonly referred to as the A+B bidding method, involves time, with an associated cost, in the low bid determination. Under the A+B bidding method, each bid submitted consists of two components:

The "A" component is the traditional bid for the contract items and is the dollar amount for all work to be performed under the contract. The "B" component is a "bid" of the total number of calendar days required to complete the project, as estimated by the bidder (calendar days are used to avoid potential misinterpretations which may arise if work days were used). The "B" component is multiplied by a factor "x" which is generally equal to or less than the estimated road user cost per day. The bid for award consideration is based on a combination of the bid for the contract items and the associated cost of the time, according to the formula:

(A)* + *(B)* x *(x)

This formula is only used to determine the lowest bid for award and is not used to determine payment to the contractor.

The contract incorporates an incentive / disincentive provision based on the "x" factor that assesses a disincentive to discourage the contractor from overrunning the time "bid" for the project and provides an incentive for early completion.

For critical projects that have high road user delay impacts, the A+B bidding method can be an effective technique to significantly reduce these impacts.

b. Design-Build

The design-build concept allows the contractor maximum flexibility for innovation in the selection of design, materials and construction methods. With design-build procurement, the contracting agency identifies the end result parameters and establishes the design criteria. The prospective bidders then develop design proposals that optimize their construction abilities. The submitted proposals may be rated by the contracting agency on such criteria as design quality, timeliness, management capability and cost, and these criteria may be used to adjust the bids for the purpose of awarding the contract.

By allowing the contractor to optimize its work force, equipment, and scheduling, the design-build concept offers greater opportunities for innovation. However, along with the increased flexibility, the contractor must also assume greater responsibility. Extended liability insurance or warranty clauses may be used to ensure that the finished product will perform as required. From the contracting agency's perspective, the potential time savings are a significant benefit. Since the design and construction are performed through one procurement, construction can begin before all design details are finalized. For example, pile driving could begin while bridge lighting is still being designed. Because both design and construction are performed under the same contract, claims for design errors or construction delays due to design errors are not allowed, and the potential for other types of claims is greatly reduced. Additionally, design-build contracts may be awarded prior to the completion of the NEPA process, which will enable the State to have the contract in place once authorized to proceed with construction.

The design-build method of contracting provides an alternative to the traditional design-bid-build contracting method, but it should only be used for projects that fit the design-build process. The contracting agency must adequately define the scope of work prior to the request for proposals. A design-build project should have sufficient scope and complexity to allow for a strong creative design component. Relatively small or simple projects such as roadway resurfacing or minor roadway widening projects do not provide significant design components, and are not ideal projects for design-build. The design-build method assists in expediting project delivery. It is not intended to provide a means for quick obligation of funds or to compensate for insufficient State personnel resources.

Federal-aid funds may participate in design-build contracts when approved and awarded using the procedures in 23 CFR Part 636.

c. Abbreviated Plans and Shortened Advertisement Period for Bids

Pursuant to 23 CFR 635.113(b), the FHWA Division Administrator may approve abbreviated plans, provided all essential information necessary to describe the work to be accomplished and to determine the reasonableness of unit prices for contract or force account work have been provided. Also, the time period for advertisement of bids may be shortened; however, a State may also need to suspend its own rules and regulations covering advertisement periods.

d. Short List of Qualified Contractors

Another technique that has been used successfully to accelerate contract bidding and award involves using a short list of qualified contractors to bid on a project. For example, a minimum of three bidders may be selected based on the following: early willingness to respond, type of work, prior demonstrated ability to move swiftly, availability, staff and equipment, and having previously worked in the area. Generally, a contractor awarded a contract as low bidder on one project is not included in the short list of qualified contractors for the next project; however, the unsuccessful bidders are.

4. Contract Requirements

Contracts for both permanent repair work and emergency repairs must incorporate all applicable federal requirements. As such, FHWA Form 1273 must be included in all contracts pursuant to 23 CFR 633.102. FHWA Form 1273 includes, but is not limited to, the following requirements:

a. Davis-Bacon Act

Generally, 23 USC. 113 requires that all laborers and mechanics employed for construction work on Federal-aid highways shall be paid wages at rates not less than those prevailing wages as determined by the Secretary of Labor under the Davis-Bacon Act. This provision cannot be waived by the FHWA. Davis-Bacon Act requirements may be waived only by executive order of the President, ref. 40 USC. 276a-5, which states, *"In the event of national emergency the president is authorized to suspend the provisions of 276a to 276a-5 of this title."*

Contract work for emergency repairs: All contract work for emergency repairs performed by contractors or subcontractors within the right-of-way of a Federal-aid highway is covered by 23 USC. 113 requirements. While contracting agencies are empowered to begin emergency repairs immediately, they must comply with 23 USC. 113 requirements so that properly documented costs will be eligible for reimbursement once the FHWA Division Administrator makes a finding that the disaster is eligible for emergency relief funding.

Contract work for debris removal only: 23 USC. 113 requirements do not apply where emergency contract work is *only* for the removal of debris and related clean up, which is not considered to be a "construction" activity for the purposes of 23 USC. 113.

However, debris removal performed in conjunction with construction, alteration, and repair work (such as highway resurfacing, re-grading, significant earthmoving, bridge repairs, etc.) is covered by 23 USC. 113.

Work by public agency forces: 23 USC. 113 requirements do not apply to State or local government agency employees who perform emergency repairs or construction work on a force account basis because government agencies (such as States or their subdivisions) are not considered contractors or subcontractors. See 29 CFR 5.2 (h). However, 23 USC. 113 requirements do apply to contracts let by State or local government agencies using an alternative procurement procedure that has been approved through the force account approval process.

b. Buy America

The FHWA's "Buy America" regulations (23 CFR Part 635.410) apply to all Federal-aid highway construction projects that permanently incorporate either iron or steel. A State may request that these provisions be waived if *"the application of those provisions would be inconsistent with the public interest" [23 CFR 635.4109(c)(1)(i)].*

c. Disadvantaged Business Enterprises (DBE)

The normal DBE requirements are applicable to all ER funded projects.

d. Americans with Disabilities Act (ADA)

The FHWA operates under the ADA regulations issued by the Department of Justice (DOJ).

According to DOJ, no waivers from these regulations are possible. The governing statute and DOJ regulations make no provision or exception for emergency relief situations. The ADA accessibility guidelines issued by DOJ, however, do provide guidance concerning temporary structures.

e. Equal Employment Opportunity (EEO)

EEO requirements prohibit discrimination and requires contractors to take affirmative action to assure equal opportunity as set forth under laws, executive orders, rules, regulations (28 CFR 35, 29 CFR 1630 and 41 CFR 60) and orders of the Secretary of Labor as modified by the provisions prescribed herein, and imposed pursuant to 23 USC. 140 shall constitute the EEO and specific affirmative action standards for the contractor's project activities under this contract.

f. Convict Labor

The convict labor prohibition in 23 USC. 114 applies to ER projects. Convict labor cannot be used in ER construction projects.

g. Use of Suspended or Debarred Contractors

Recipients of Federal funds are prohibited from doing business with contractors who have been suspended or debarred. This prohibition includes contractors with principals who have been suspended or debarred. Recipients are responsible for determining whether any contractor or its principals are suspended or debarred. In addition to certifications provided in

FHWA Form 1273, recipients should check the Excluded Parties List System that is maintained by General Services Administration at: https://www.epls. gov /.

N. Environmental Considerations:

Repair projects under the ER program must comply with the requirements of the National Environmental Policy Act (NEPA) of 1969. Emergency repairs to restore essential travel, minimize the extent of damage, or protect remaining facilities are normally classified as categorical exclusions under 23 CFR 771.117(c)(9), as are ER projects to restore permanently the existing facility in-kind at the existing location, ref. 23 CFR Part 771.117(d). However, if impacts to protected or otherwise sensitive or high-value resources are possible, advance coordination with the appropriate local, State, and Federal resource agencies should be closely considered to avoid or minimize project delays or shutdowns. On occasion, an ER project that includes a betterment, whether or not eligible for ER funding, may require further NEPA review. Although on the surface a project may appear to qualify for a categorical exclusion, certain betterments may need either an environmental assessment (EA) to determine whether or not the project will cause significant environmental impacts, or an environmental impact statement (EIS) if significant impacts are predicted. This is illustrated by the following example:

Project Betterment Requiring Environmental Evaluation

When repairing a section of roadway inundated and seriously damaged by floodwaters, it was determined that a grade raise could be economically justified for ER funding. Raising the grade of the roadway will require small amounts of additional right-of-way from adjacent wetland areas. In addition, in future flood events, the higher roadway grade could impound additional water and flood other upstream areas. As a result of the project's potential impact on wetlands and future flooding patterns, further evaluation was necessary to determine the appropriate level of NEPA documentation.

The NEPA project development process provides the final Federal-aid highway project decision, occasionally including a facility on new location. As noted above, ER projects to construct replacement facilities may require environmental assessments or environmental impact statements, depending on the potential level of impacts to resources, the value of the resources, and what, if any, legal protections apply to the resources.

However, even replacement facilities constructed at the existing location of the damaged facility may require extra environmental evaluation beyond that needed for a routine categorical exclusion. These situations are illustrated by the following examples:

Replacement at New Location

A roadway was permanently submerged by water backing up behind a naturally created dam, and it has been determined replacement of the inundated highway facility at its existing location is neither practical nor feasible, and various alternate locations may be available to relocate this section of highway.

The NEPA process documents consideration of appropriate project alternatives and their potential impacts and determines that the preferred alternative is replacement of the old facility on a specific new location or site. Although a categorical exclusion can be used if circumstances merit, early environmental coordination may determine that an EA or an EIS is necessary to do this.

Replacement at Existing Location

An existing bridge over a river has been damaged beyond repair but can be replaced with a bridge of comparable width and length at the same location. However, this section of river contains critical habitat for a federally listed endangered species, which would be seriously impacted during the scheduled construction period.

As a result of this potential impact, the project decision could not be categorically excluded, and additional NEPA evaluation and documentation was necessary.

O. Design Standards

Reconstruction of damaged roadway and bridge facilities must meet adequate standards, including appropriate safety features. Reconstruction of extensively damaged facilities, including betterment projects when adequately justified, should meet the current applicable design standards. Replacement of roadway facilities other than bridges is limited to the existing number of lanes and surface type.

Bridges may be replaced with a facility that meets current geometric and construction standards required for the type and volumes of traffic that such a facility will carry over its design life.

P. State Emergency Manual

An important element of emergency procedures is the State's emergency operation plan. Since Federal-aid funds are not available for maintenance and State maintenance personnel will handle much of the emergency operations, the State's instructions should be carefully reviewed.

It may be appropriate to supplement the State's instructions with portions of this manual, for example, Chapter V, Disaster Assessment and Damage Survey Summary Report and Chapter II, Eligibility of Damage Repair Work.

Appendix A. Sample ER Request Memorandum

Subject: [*State Name*] - Emergency Relief (ER) Event,
[*Event Title, Disaster Code*]

From: [*Division Administrator Name*]
Division Administrator
[*City, State*]

In Reply
Refer To:

To: [*Office of Program Administration Director Name*]
Director, Office of Program Administration

The purpose of this memorandum is to advise you that the [*State*] Division Office has approved the subject event as eligible for ER funding. In accordance with the Delegation of ER "Finding" Approval memorandum, dated June 28, 1999, we are providing the following information:

1. [*Example event description*] Beginning on March 10, 2009, severe storms caused flooding and wind damage resulting in debris and roadway damages to Federal-aid highways in the southwestern region of the State. Some areas received up to 15 inches of rainfall over a 48 hour period. Several major roadways, including State Highway 10 and Interstate Highway 90 were severely damaged and were closed to traffic due to flood damages. Typical damages include debris deposits on many roadways, shoulder washouts, pavement damage, sign and signal damage, and scour around bridge piers.

2. The disaster number for this event is ________.

3. The total cost of eligible damage is $__________ with a Federal share of $___________.

4. Counties with eligible damage sites are ______________________________.

5. The following U.S. Congressional Districts were affected: __________________.

We request that ER funding in the amount of $_____________ be made available for the subject disaster. If you have any questions concerning this request, please contact ________________.

Appendix B. Sample State Letter of Intent

TO:_______________________________________, Division Administrator
Federal Highway Administration

FROM:____
Administrator and State Transportation Engineer

SUBJECT: [Severe Flooding in] [State]
[Date]

Dear ________________

Under the provisions of Title 23, U.S.C., Section 125, this is notice of intent by the [State] Department of Transportation to request emergency relief funds to assist in the cost of repairing damages on the Federal-aid highways in [State] caused by the extreme runoff and flooding following the [storm] beginning [Date].

Attached is a copy of the Declaration by Governor ____________ of a State of Disaster in [State] on [Date].

Preliminary estimates of the damages sustained to the Federal-aid highways will be forwarded within a few days when assembled.

We are proceeding expeditiously to maintain two-way traffic at all locations and to repair those sections sufficiently to protect facilities from further damage.

Yours Sincerely,

____

Attachment

Appendix C. Sample FHWA Acknowledgement Letter

Name
Title
State Department of Transportation

Dear __________:

This is to acknowledge receipt of your letter of intent, dated ______, to request Emergency Relief Funds, authorized under 125 of Title 23, U.S.C., for the repair of damage to Federal-aid highways resulting from the ___________ (event) of __________(date).

You should proceed with performance of emergency operations, including emergency repairs, on the Federal-aid highways necessary to restore essential traffic, to protect the remaining facilities, and to reduce the extent of damage. Also, you may proceed with preliminary engineering, meaning surveys, design, and preparation of construction plans, to perform the permanent restoration work required as an associated part of the emergency operations, and to use State forces and/or negotiated equipment rental contracts as necessary to perform the work.

The eligibility of such work for ER funds will be contingent upon a favorable finding by the FHWA Division Administrator, on the eligibility of the disaster, and subsequent approval of the work by FHWA.

The basis for the Division Administrator's decision will be the Damage Survey Summary Report, which must be submitted to this office. The DSSR, among other requirements, must include estimates of cost to repair and reconstruct the damaged Federal-aid highways.

My office will be meeting [has met] with members of your staff to make arrangements for reviewing the disaster damage and assisting in preparing the Damage Survey Summary Report and site damage reports. The Damage Survey Summary Report is to be submitted within --- weeks, if possible. If additional time is required, please advise, including the reasons why the extra time is necessary.

If FHWA concurs in the disaster, all emergency work must be included in a program of emergency repair projects. The program, when submitted for approval, shall include a detailed outline of the necessary emergency operations performed and a description of the permanent restoration work proposed. Permanent restoration work other than that performed as an associated part of the emergency operations, shall not be performed prior to program approval and authorization by this office.

Sufficient record keeping must be done to permit audit of costs on a site-by-site basis.

Sincerely yours, [Division Administrator]

APPENDIX D. SAMPLE GOVERNOR'S PROCLAMATION

Floods and rapid runoff, commencing on [date], were experienced throughout the following counties in [State] as a result of extremely heavy rains: [list of counties]. The flooding and associated runoff have produced serious and extensive damage to both private and public property. As a consequence, this State has sustained severe damage to its road systems, which include bridges, roadbeds, and other facilities. Damage occurred on Federal-aid highways.

Damage throughout the [region] part of the State has been of such an extent that immediate repairs have been necessary. Such conditions constitute an emergency as is contemplated by the terms of Sections 125 and 120(e) of Title 23, U.S.C.

Therefore, I ___________, Governor of the State of _________ do hereby proclaim an emergency to exist throughout the [region] of the State as a result of flooding and runoff conditions and consequent danger to life and damage to property including Federal-aid highways.

The immediate repair and reconstruction of the damaged highways is vital to the security, well-being, and health of the citizens of the State of (State); and the Federal Highway Division Administrator is hereby requested to concur in the declaration of this emergency.

In testimony whereof, I have hereunto subscribed my name and caused the seal of the State of ________ to be affixed at [City], the ___ day of [month], A.D. [year].

Governor of ___________

ATTEST:

____

Secretary of State

APPENDIX E. DETAILED DAMAGE INSPECTION REPORT

U.S. Department of Transportation
Federal Highway Administration

DETAILED DAMAGE INSPECTION REPORT

(Title 23, Federal-aid Highways)

Report Number	
Sheet	______ of ______
Location *(Name of Road and Milepost)*	FHWA Disaster Number
	Inspection Date
Description of Damage	Federal-aid Route Number
	State / County

Cost Estimate

	Description of Work to Date (Equipment, Labor, and Materials)	Unit	Unit Price	Quantity	Cost	
					Completed	Remaining
Emergency Repair		ea				
		ea				
		ea				
		ea				
		ea				
		ea				
		ea				
		ea				
		ea				
		ea				
		ea				
		ea				
	Method			Subtotal		
	☐ Local Forces ☐ State Forces ☐ Contract			PE/CE		
				Emergency Repair Total		
Permanent Restoration		ea				
		ea				
		ea				
		ea				
		ea				
		ea				
		ea				
		ea				
		ea				
		ea				
		ea				
		ea				
	Method				Subtotal	
	☐ Local Forces ☐ State Forces ☐ Contract				PE/CE	
					Right-of-Way	
					Perm. Repair Totals	

Environmental Assessment Recommendation ☐ Categorical Exclusion ☐ EA/EIS	Estimated Total	
Recommendation ☐ Eligible ☐ Ineligible	FHWA Engineer	Date
Concurrence ☐ Yes ☐ No	State Engineer	Date
Concurrence ☐ Yes ☐ No	Local Agency Representative	Date

Form FHWA-1547 (Rev. 4-98)

Appendix F. Sample State Letter for Quick Release of ER Funds

Name __________________
FHWA Division Administrator

Dear _____________:

We appreciate your timely acknowledgement letter of ___[date]___ sent in response to our letter of intent dated ___[date]____. We are continuing to conduct necessary emergency operations and repairs to maintain traffic throughout the disaster area.

We are performing damage assessments throughout the area. At this time it is not possible to provide exact cost estimates; however, damage to Federal-aid highways is anticipated to exceed ___[amount]___. We expect to be able to provide a more accurate estimate [within "x" weeks].

Our budget for emergencies is limited and local governments have even greater financial constraints with limited cash flows available to fund emergencies. Consequently, we are requesting approval of ER funding for this disaster with a quick release of emergency relief funds to allow us to proceed expeditiously with emergency repairs to Federal-aid highways.

We are requesting a quick release of ___[amount]___ for these emergency repairs. Additional allocations will be requested as damage survey teams inventory damage.

Sincerely yours,

____
(State Department of Transportation Official)

Date: ______________________

Appendix G. Sample State Letter Requesting ER Funds

Date ________

To: FHWA Division Administrator

Subject: Heavy Rains and Flooding March 23-26, 2009

Dear ___________

Under the provisions of Title 23, U.S.C. Section 125, the (State) Department of Transportation is requesting Emergency Relief (ER)Funds to assist in the cost of repairing damages on Federal-aid highways in (State) damaged as a result of extreme runoff and flooding following the heavy rains beginning March 23, 2008. Based on the preliminary estimate of damage, we are requesting ($3) million in ER funds. Following a favorable determination of eligibility for ER funds by the Federal Highway Division Administrator, we will submit detailed damage inspection report with the scope of work and cost estimate for each site as a program of projects.

Sincerely

____

Title
[State] DOT

In: Emergency Relief Programs for U.S. Highways and Roads ISBN: 978-1-61942-858-4
Editors: Cody O. Harrison and Ted J. Gibson © 2012 Nova Science Publishers, Inc.

Chapter 2

EMERGENCY RELIEF FOR FEDERALLY OWNED ROADS: DISASTER ASSISTANCE MANUAL*

The United States Department of Transportation

GETTING STARTED

* This is an edited, reformatted and augmented version of The United States Department of Transportation, Federal Highway Administration publication No.: FHWA-FLH-11-001, dated April 2011.

1. INTRODUCTION

The *Emergency Relief for Federally Owned Roads Program*, or ERFO Program, was established to assist Federal agencies with the repair or reconstruction of Federal roads, which are found to have suffered serious damage by a natural disaster over a wide area or by a catastrophic failure (23CFR668.201).

1.1. Purpose

The purpose of this manual is to provide federal land management agencies with guidance and instructions to apply for federal assistance under the ERFO program. Federal, tribal, state, and local governments that have the authority to repair or reconstruct federal roads may apply for ERFO funds, but only the federal land management agencies (FLMA) can apply directly as an "Applicant." The other governmental entities must apply through an "Applicant."

1.2. Program Intent

The intent of the ERFO program is to pay the unusually heavy expenses in the repair and reconstruction of Federal roads 23CFR668.205 (a). The ERFO program is *not* intended to cover all repair costs nor interim emergency repair costs that are necessary to repair or reconstruct Federal roads.

Agencies have the responsibility to perform emergency repairs and fund the unexpected expenditures, shift project priorities and manage reduced traffic service levels that a natural disaster can present. Emergency relief work shall be given prompt attention and priority over non-emergency work.

1.3. Authority

The Emergency Relief for Federally Owned Roads Program is authorized under Title 23, United State Code, Sections 120 and 125.

This law authorizes an emergency fund and expenditure for the repair or reconstruction of roads and trails that the Secretary of Transportation has found to have suffered serious damage as the result of natural disasters over a wide area or as the result of catastrophic failures.

1.4. Stewardship and Oversight

The Office of Federal Lands Highway is responsible for efficient and effective management of public funds entrusted by Congress and for ensuring that the ERFO Program is administered consistent with laws, regulations, and policies.

The applicants are expected to prioritize projects for repair that are in the public best interest based on available funds.

1.5. Overview

Federal, tribal, state, and local governments must be prepared to respond quickly when disasters strike. The purpose of this manual is to provide the applicants with procedural information to enable this quick response.

Section 1 presents a brief overview of the ERFO program, its statutory authority and intent.

Section 2 presents the ERFO Program's administrative process from disaster approval to disaster closeout.

Section 3 presents the funding process beginning with requesting funds through disaster closeout and includes accountability and reporting requirements.

Section 4 identifies ERFO eligible roads and eligible repairs.

Section 5 lists roles and responsibilities of individuals within the agencies and Federal Lands Highway offices.

1.6. Highlights of Federal Emergency Relief Programs

The federal government has three emergency relief programs which provide assistance for the repair and reconstruction of seriously damaged roads and trails caused by natural disasters or catastrophic events.

1.6.1. ERFO Program

a) Authorized under Title 23, United States Code, Sections 120 and 125.
b) Provides financial assistance to approved Applicants and eligible roads and trails that have been defined as federal roads in Title 23, Code of Federal Regulations (CFR).
c) Funds are provided from the Highway Trust and occasionally the General Fund through the Federal-aid Highway Emergency Relief Program.
d) Further clarified under CFR, Part 668; ERFO Disaster Assistance Manual; and with interagency agreements.
e) The federal share payable under this program is 100 percent of the cost.

1.6.2. Federal-aid Highway Emergency Relief (ER) Program

a) Authorized under Title 23, United States Code, Sections 120 and 125.
b) Provides assistance for roads that have been defined under Title 23, CFR, Part 668 as federal-aid highways (all state and local public roads that have been classified or designated as Interstate Highways, National Highway System routes, rural and urban arterials, and major collector roads).

c) Local streets and rural minor collectors are not eligible.
d) A cost-sharing program established by law between the federal government and the state transportation departments and/or local highway authorities. The federal share ranges from 80 percent to 100 percent.
e) Some federal roads are also federal-aid highways. The state transportation department and/or local highway authorities may elect to receive reimbursement for eligible emergency repairs from either the ER or ERFO programs for these roads.
f) The FHWA published an Emergency Relief Manual, which includes detailed guidance on the Emergency Relief process, including the eligibility of repair work for funding.

1.6.3. Robert T. Stafford Disaster Relief and Emergency Assistance Act, Federal Emergency Management Agency

a) Authorized under Title 42, United States Code, Section 5121, et seq.
b) Grants and eligibility of repairs are determined by individual presidential declarations that may authorize state or local government for the repair, restoration, reconstruction, or replacement of public facilities that are damaged or destroyed by a major disaster and for associated expenses incurred by such government.
c) Considered as an alternative funding source to repair damage that is ineligible under the ERFO or Emergency Relief programs for state, county, and tribal governments.

For additional information on the Federal Emergency Management Agency (FEMA) Program, contact the FEMA Office of Emergency Information and Public Affairs in Washington, D.C. A copy of its Public Assistance Guide, FEMA 322, may be obtained from FEMA, RR-IS-EN, 500 C Street SW, Washington, D.C. 20472 or over the internet at http://www. fema.gov/.

1.7. Federal Share Payable

The federal share payable for repair or reconstruction of forest highways, forest development roads and trails, park roads and trails, parkways, public lands highways, public lands development roads and trails, [refuge roads and trails], and Indian reservation roads is 100 percent. The total cost of a repair may not exceed the cost of repair or reconstruction of a comparable facility (U.S.C. Title 23, Sec. 120 Federal share payable).

1.8. Funding

Emergency funds are appropriated annually from the Highway Trust Fund. The emergency funds may be supplemented by special appropriations out of the Highway Trust Fund or the General Fund.

During periods of insufficient funds, the agencies should continue to make necessary repairs using their funds.

Other funds obligated shall be reimbursed when ERFO funds are available (23 U.S.C. Section 125(c)).

a) When ERFO funds are insufficient to meet anticipated needs, the FLH will place a priority on allocating funds for:
   1) Reimbursement (payback) of Applicant funds, including state, county, local and tribal government funds used on approved ERFO repairs;
   2) Reimbursement of FLHP funds (FH, PRP, IRR, RR, and PLH-D) used on approved ERFO repairs;
   3) Construction engineering and contract modifications for ongoing ERFO construction projects;
   4) Emergency repairs of high-use facilities that are critical to the agencies' mission and/or health and welfare of the local community/region; and
   5) Fiscal year disasters.

b) ERFO funds will not be approved to supplant other funds for the correction of pre-existing, non-disaster related deficiencies nor will ERFO funds be approved for the repair or reconstruction of facilities affected by long term, pre-existing conditions, or predictable developing situations (23CFR668.105(b) and 23CFR668.109(c)(6)).
c) ERFO funds shall not duplicate assistance under another Federal program or compensation from insurance or any other source. Diligent efforts shall be made to recover costs from legally responsible parties prior to applying for ERFO funds.
d) ERFO funds will not be approved for heavy maintenance, routine emergency repair activities, emergency services (e.g. ambulances, helicopters), repair or relocation of utilities, snow removal, material, equipment, or economic losses (e.g., stockpiled material, maintenance equipment, lost revenue, or personal injury).
e) While ERFO funds may be used to restore federal roads and bridges to pre-disaster conditions, agencies may supplement ERFO funds to address agency priorities not approved for ERFO funding.

1.9. Funding Threshold

a) The minimum threshold for federal road and bridge damage is $700,000 for each natural disaster and catastrophic failure.
b) The minimum threshold for each damage site (Emergency and Permanent Repairs) is $5,000.
c) If the combined road and bridge damages for an individual natural disaster for all the agencies that manage eligible federal roads is less than $700,000, the agencies are expected to fund the repair costs using their agency funds.

1.10. Environmental Compliance

The Applicant shall comply with all pertinent federal, state, and local laws.

1.11. Economic Hardship

Agencies shall strive to anticipate their probable needs for additional funding that may result from disaster events and budget accordingly. Economic hardship is not a factor in determining eligibility.

1.12. Approvals/Determinations

a) The Federal Lands Highway (FLH) Division Engineer (DE) determines if a natural disaster or catastrophic failure is eligible for ERFO funding.
b) The finding for natural disasters will be based on both the extraordinary character of the natural disturbance and the wide area of impact. Storms of unusual intensity occurring over a small area do not meet these conditions.
c) FLH DE determination concerning the extraordinary nature of the natural disturbance is not necessary when the President issues a Major Disaster Declaration under the Robert T. Stafford Disaster Relief and Emergency Assistance Act (42 U.S.C. 5121 et seq.).
d) Emergency repairs do not require prior concurrence of the FLH DE. It is incumbent on the agency with jurisdiction over the road to protect remaining facilities, minimize damage, and restore essential travel. However, sufficient documentation, including pictures of the damage must be provided to establish eligibility.
e) Permanent repairs require prior program approval unless they are necessary and practical to accomplish as emergency repairs.

1.13. Timing

a) Emergency relief projects shall be promptly constructed and given priority over non-emergency work. (23CFR668.205 (b) and (e)).
b) Projects not under construction by the end of the second fiscal year following the year in which the disaster occurred will be reevaluated and will be withdrawn from the approved program of projects unless suitable justification is provided by the agencies to warrant retention 23CFR668.205 (e).

1.14. Interagency Agreement Needed for Participation in the ERFO Program

Participation in the ERFO program requires an interagency agreement be in place prior to release of funds.

2. The Process

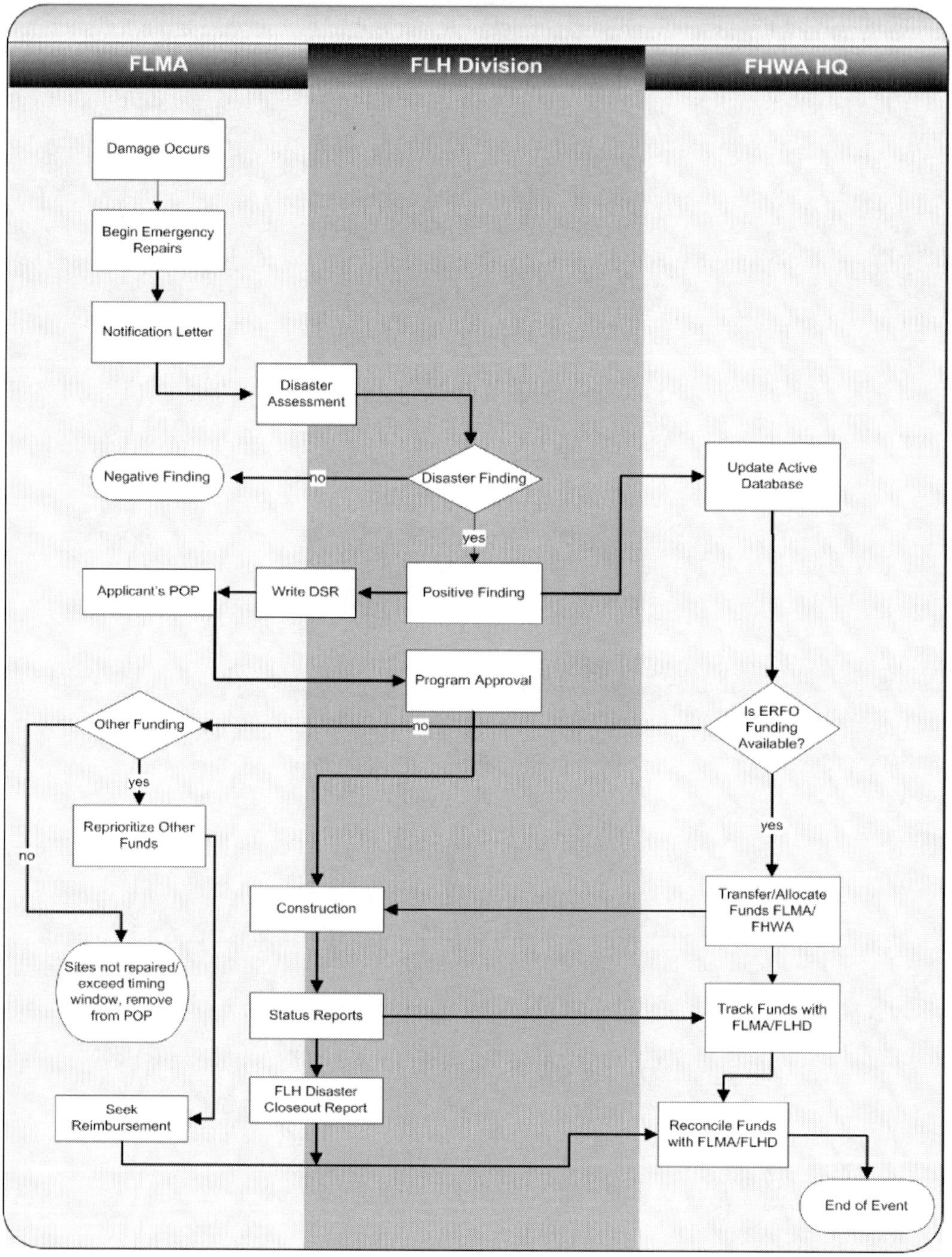

ERFO PROCESS TIMELINE

Disaster Occurs

↓

Notice of Intent (NOI): FLMA submits by letter to FLHDE **within 45 days**

↓

Acknowledge NOI: FLH DE replies by letter **within 14 calendar days**

↓

Disaster Assessment: FLHD completes **as soon as practical**

↓

Disaster Assessment Report: FLHD completes **ASAP** following Disaster Assessment

↓

Positive/Negative Finding: FLH DE replies by letter **within 6 weeks** after the receipt of the NOI

↓

Program of Projects (POP) Request: FLMA submits **within 3 months** after the date of the Positive Finding

↓

POP Approval: FLH DE completes **within 30** calendar days

↓

Funds Request: FLMA submits for current fiscal year obligations

↓

Emergency Repairs: FLMA completes **within 180 days** of the damage

↓

Permanent Repairs: FLMA must award construction contract by the **end of the second fiscal year** following the fiscal year in which the disaster or catastrophic failure occurred. This allows from **25 to 36 months** completing design and beginning construction. Projects may be withdrawn from the POP unless a time extension is approved by the FLH DE

↓

Time Extension Justification: FLMA must submit to the FLH DE for approval at least **60 days prior** to the end of the second fiscal year following the fiscal year in which the disaster occurred

↓

Disaster Status Report: FLMA submits **by November 30** for the preceding fiscal year

↓

Prior-year Unobligated Balance Report: FLMA submits to FLH headquarters within the first quarter of the new fiscal year

↓

Annual ERFO Funds Request: FLMA submits needs for the next fiscal year to the FLH ERFO Program Manager **by September 15**

↓

Final Disaster Status Report: FLMA submits the final closeout report to the FLH DE within 6 months after the final inspection

↓

Project Closeout/Funds Reconciliation: FLMA completes permanent repairs and closes out the disaster within 4 years of the disaster

2.1. Damage/Disaster

The Federal Highway Administration (FHWA) will determine if there has been a natural disaster (i.e., severe storms, tidal waves, earthquakes, landslides) of sufficient extent and intensity to warrant ERFO funding consideration. To this end, FHWA defines the following;

Natural disaster: An unusual natural occurrence such as a flood, hurricane, severe storm, tidal wave, earthquake, or landslide which causes serious damage.

Serious damage: heavy, major, or unusual damage to a highway which severely impairs the safety, or usefulness of the highway or results in road closure and must be beyond the scope of heavy maintenance.

Wide area: a tract sufficiently large to encompass parts or all of several entities such as counties, states, or federal agency management units (individual forests, parks, refuges, regions, districts, and so on) and can include areas outside of federal lands. Significant flooding would involve a number of major drainage basins.

Special Note: Serious damage to roads caused by extraordinary storms occurring over small areas would not be considered eligible for ERFO funding.

Catastrophic failure: The sudden failure of a major element or segment of the highway system due to an external cause. The failure must not be primarily attributable to gradual and progressive deterioration or lack of proper maintenance. The closure of a facility because of imminent danger of collapse is not in itself a sudden failure.

2.1.1. Occurrence

a) When a natural disaster or catastrophic failure occurs (an "event"), an *Informal Notice of Intent* shall be initiated by the Applicant within two weeks via telephone or email to the Division ERFO Coordinator.
b) It is the responsibility of the Applicant to assess the damage no more than 45 days after the event and to submit a *Formal Notice of Intent* to the Federal Lands Highway DE. See Appendix C.1 for a sample letter and C.1.1 for letter instructions.
c) The DE will respond within two weeks of receipt of the Applicant's formal Notice of Intent letter with an *Acknowledgment Letter* (see Appendix C.2).

Special Notes:

- Photograph damage before beginning emergency repairs. Eligibility of emergency repairs will be based on submitted photographs.
- The Applicant may request Accelerated Release of ERFO funds for emergency repairs prior to the approval of ERFO POP. These funds are only to be used by agencies to make emergency repairs to re-establish their vital transportation system(s) after an event. A request for Accelerated Release shall be reviewed by the FLHD on a case-by-case basis (see_Sections 3.4 and Appendix C.6).

2.1.2. Disaster Assessment

The ERFO Coordinators for the FLHD and the Applicant must plan and conduct an assessment of the damage and prepare a Disaster Assessment Report. This report shall document the cause and extent of damage and provide a recommendation to the DE for either a positive or negative finding. If a Presidential Disaster Declaration has been issued, the DE may waive or limit the assessment efforts.

Isolated damage sites unrelated to, or away from, the eligible disaster may be excluded from the Disaster area.

Special Note: Two disasters treated as one. Disasters can occur so closely in time that the damage to roads and trails is difficult or impossible to evaluate and track as two separate events. In these cases, the Applicant should request in a letter to the DE that the two disasters be recognized as a single event. Treating two disasters as one may simplify the processing of the repair and reconstruction projects by having a single ERFO finding resulting from the two disasters. This may not be desirable if one or both of the disasters potentially are subject to the $100,000,000 cap per disaster per state. The supporting data must prove that facilities have suffered severe damage in both disasters and such damage is of sufficient severity to justify ERFO assistance.

a) Applicant Responsibility Checklist. The Applicant must name a representative who is responsible for providing the following:
   1) In the case of catastrophic failure, obtain details of the cause of the failure, responsible parties, and impact of the damage;
   2) To expedite the disaster assessment, locate and identify damage sites prior to field assessment;
   3) Current bridge inspection reports to assist with the identification of bridge damage caused by the event;
   4) Maps of the road system and identify eligible roads; and
   5) Available pertinent data on the cause and extent of the damage such as:
      i) Rainfall,
      ii) Runoff data (hydrograph),
      iii) Dates of disaster occurrence,
      iv) Parties who may be responsible,
      v) Impact of the damage sites and/or facilities,
      vi) Information regarding the damage,
      vii) General maps of the damaged area, and
      viii) Photographs of specific damage in each area.
      ix) Cost Documentation for any emergency repairs.

Special Note: The ERFO Coordinator will not conduct a field assessment until all of the information is provided. Delays in submitting information will not extend the eligibility time period.

b) Disaster Assessment – Field Review

The Team will review a representative sample of damage sites in each area cited by the Applicant in the Notice of Intent letter to determine the extent, type, cost, impact, and eligibility of the damage.

In addition, the Team will develop a preliminary estimate of costs to repair damaged sites and/or facilities to their pre-disaster conditions.

c) Disaster Assessment – Report to DE

The ERFO Coordinator will prepare the *Disaster Assessment Report* for submission to the DE.

The report must either support or dispute eligibility for ERFO funding and provide a recommendation of either a positive or negative finding (see Appendix C.3).

The report shall include the following elements:

1) Natural disaster intensity and severity; In the case of flooding or severe storms, the report shall include available information regarding high-water elevations, stream-gage data, and rain-gage data. This information is not required if the President issues a *major disaster declaration*;
2) Catastrophic failure cause and unexpected nature. Document the significance of the damaged facility, the impact to the road users, and the sudden nature of the failure. Explain if the failure was caused by the expected gradual and progressive deterioration of materials or by lack of maintenance. If appropriate, give the names of parties that may have contributed to the failure. Also state if compensation will pay for any of the repairs, such as insurance or cost-share;
3) Dates of the occurrence;
4) Extent of the affected area, including areas outside the lands administered by federal agencies;
5) Description of the damage to federal roads and other roads (federal-aid highways, state, county, and city);
6) Copies of emergency relief reports and newspaper articles, as available;
7) Copies of presidential disaster declarations and/or governor proclamations;
8) Approximate cost of repairs or reconstruction of the damage;
9) Photographs of typical damage and appropriate maps showing the location of each area and extent of damage;
10) Copies of ERFO coordinators' trip reports;
11) Discussion of the impact of the subject damage. For example, a destroyed bridge with limited traffic has little impact, but a bridge serving as theprincipal access to homes, large tracts of land, or carrying large amounts of traffic will have a significant impact and therefore, may qualify for Accelerated Release of ERFO funds for emergency repairs; and
12) Concluding statement. A final summary statement indicating that a natural disaster has (or has not) caused serious and widespread damage to federally-owned roads or a catastrophic failure has (or has not) occurred resulting in a disastrous impact to access to transportation services. Include a specific recommendation for a finding.

2.2. Disaster Finding

The DE will make a disaster finding determination no later than six weeks after receipt of the Notice of Intent.

Positive Finding Determination The DE will issue a letter to the Applicant committing ERFO funds to repair eligible damage, subject to availability. The positive finding letter will establish the disaster number (see Appendix C.4).

Negative Finding Determination The DE will advise the Applicant by letter when the damage is not eligible for the ERFO program (see Appendix C.5).

Special Note: Eligibility questions that cannot be resolved by the FLHD and the Applicant may be appealed through the Office of Federal Lands Highway Associate Administrator with copies to the Director, Office of Program Development, and the ERFO Program Manager. The appeal should be submitted within two weeks of the receipt of the negative finding.

2.3. Damage Survey Report (DSR)

For disasters with few sites, the DSR(s) may be prepared while making the disaster assessment. The FLHD ERFO Coordinator and Applicant will jointly develop a plan to proceed with DSR preparation. DSRs should be prepared and submitted for approval as soon as possible after the positive finding. In some cases, this is not always possible. DSRs must be completed and submitted for approval within nine months of the positive finding.

2.3.1. DSR Preparation

The Damage Site Survey Team shall make a detailed inspection of, and prepare a DSR for, each damage site. Every report should include a description of the extent and cause of the damage, as well as a proposal for a repair plan and cost estimate.

When completed, all team members shall sign the report. When the team consists of Applicant and FLHD personnel, the FLHD team member shall indicate if they think the site is eligible or ineligible. (For detailed instructions on how to complete a DSR, see Appendix C.9).

Special Note: The Applicant must provide two area maps complete with all damage sites clearly numbered according to the corresponding DSR number (see Appendix C.9.1 and 9.2). One map each shall accompany copies of completed DSRs to the Applicant and to the FLHD.

a) In addition to completing the DSR, the team shall identify *emergency* repairs completed and emergency repairs remaining.
b) Identify *permanent* repairs necessary to restore the site to its original condition.
c) Identify major environmental issues that may have a substantive affect on repair costs and timelines.
d) Justify betterments, if proposed.
e) Identify repair options, if applicable.
f) Identify level of environmental review expected [e.g., Categorical Exclusion (CE, Environmental Assessment/Finding of No Significant Impact (EA/FONSI), Environmental Impact Statement (EIS)].
g) Provide detailed high quality color photographs for each site properly numbered and referenced to a particular damage site and DSR.
h) Provide a sketch of damage and proposed repairs.
i) Provide quantity calculations for major work items.
j) Provide Road Management Objectives for forest development roads that include the following:
    1) Line Officer approval;

2) Design criteria (critical and design vehicles, surfacing);
3) Operation criteria (traffic service level); and
4) Maintenance criteria (future and present)

2.3.2. DSR Personnel – Roles and Responsibilities

Applicants are encouraged to proactively conduct a damage site survey and begin preparation of the DSR. This can greatly accelerate advancement of the Program of Projects. In a case where the FLHD is unable to provide team personnel, the Applicant personnel will be authorized to perform the damage site survey and complete the DSRs. See Appendix C.8 for a list of recommended equipment.

a) FLHD personnel will review all DSRs and may conduct field reviews of selected sites prior to approving the damage description, scope of the proposed repair, and cost estimate.
b) Based on the results of these reviews, the Applicant will modify the DSR as deemed necessary by the FLHD ERFO coordinator.

2.3.3. DSR Submittal

The Applicant must submit the original DSR to the DE for approval.

2.3.4. Combining Damage Sites

It may be appropriate in some situations to combine damage sites so as to prepare a single DSR for a road or road segment. Typically, the use of one DSR for combined damage sites is reserved for damage that will be repaired in the same construction contract. The team will identify each individual site by mile post and will prepare a cost estimate complete with recommended repairs and photographs for each site. The DSR form will then represent a summary of the costs to repair the damage sites along the total length of road or road segment. However, damage sites *shall not* be combined just to meet the $5,000 minimum threshold requirement for ERFO consideration.

Sites such as bridge replacements and major landslides shall have separate DSRs.

2.3.5. Amended DSRs

Amended DSRs must be completed when the following applies:

a) The Applicant wants to revise the scope of approved repairs during design or construction. Examples of scope revisions that require amended DSRs include:
   1) Modifying, adding, or deleting major construction items;
   2) Revising the roadway cross-sections;
   3) Extending the project limit; and
   4) Electing to decommission the road.
b) The cost estimate increases more than $50,000 for DSRs less than $250,000.
c) The cost estimate increases more than 20 percent for DSRs over $250,000.
d) When multiple sites have been combined into one DSR and some of the site repairs must be delayed, those deferred repairs shall be separated from the projects that can be repaired promptly and accounted for separately.

2.4. Program of Projects Proposed by the Applicant

The initial POP should be submitted to the DE within 90 days after receipt of the *positive finding*. The submittal must contain the following:

a) *POP Letter*: includes the completed DSRs and the Program of Projects Table, requests ERFO funds for the repair of damaged sites (see Appendix C.10), and requests FLHD assistance to repair damage, if applicable.
b) *POP Table:* a list of all the DSRs submitted for approval indicating the proposed Agency responsible for repair (see Appendix C.11).
c) *All DSRs*: DSRs *must be submitted within nine months* of the *positive finding* date to be eligible.

DSRs on eligible roads maintained for high clearance vehicles and administrative roads classified as public lands development roads will be approved only for cases where the Applicant funds the repair using agency funds and seeks reimbursement from the ERFO Program later.

Advance ERFO funds will not be authorized for these two categories.

2.4.1. Request FLHD to Perform Work

If the applicant wants FLHD to perform ERFO repairs, they need to make the request so there is enough time to allow contract award by the end of the second fiscal year following the fiscal year the disaster occurred. The DE shall inform the Applicant by letter within 30 calendar days of the request for work that the FLHD agrees to perform.

2.4.2. Advance Funding Limitations

Do not request advance funding for roads shown below. The applicant may only request reimbursement after repairs have been completed with other funds. Show the fund source to be reimbursed on the Proposed Program of Projects.

a) Roads maintained for high clearance vehicles.
b) Administrative roads classified as Public Lands Development Roads.

2.5. Approved Program of Projects (POP)

2.5.1. Program of Projects Approval Letter

The POP approval letter from the DE establishes the maximum amount of funds approved for the repair of eligible damage. The approved POP shall be completed within 30 calendar days of receipt of the Applicant's POP (see Appendix C.12). The purpose of this letter is to:

a) Identify projects in the Applicant's proposed programs that are approved, including any required changes.
b) Identify *future project approvals* required by the DE such as plans, specifications, and estimates; concurrence in award; rejection of bids; and project acceptance. Project

selection will be based on project complexity, cost, and past performance; it may include randomly selected projects.

c) Identify work the FLHD agrees to perform.
d) Provide approval of the use of *force account* when it is determined that it is in the best interest of the public.

2.5.2. ERFO Fund Availability

Even though the funding level is approved in the DE's Program of Projects Approval Letter, ERFO funds may or may not be available. If ERFO funds are not available, the Applicant is expected to use other funds to cover the cost of the approved repairs until ERFO funds are available for reimbursement.

2.5.3. Revisions to the POP

When DSRs are amended or new sites added, a revised POP request must be submitted to the DE for approval as soon as possible. The Agency responsible for submitting the revised POP varies as follows:

a) Applicant. When the Applicant is the lead Agency for the repair of a project, the Applicant must request a revision to the DSR and the POP to ensure the availability of funds. It requires approval by the DE.
b) FLH Division. When the FLHD is the lead Agency for the repair of a project, the DE will revise the Program of Projects with concurrence of the Applicant.

As stated previously, new DSRs will not be approved if submitted more than nine months after the date of the positive finding.

2.6. Environmental Compliance

The lead Agency must comply with all applicable federal, state, and local environmental laws and permit requirements dealing with issues such as wetlands; resource conservation; clean water; endangered species; farmland; unexploded ordnances; and cultural resources. Supporting documentation that shows compliance with all applicable federal, state, and local environmental laws must be submitted to FLHD.

2.6.1. Emergency Repairs

Regardless of which Agency administers the project, emergency repairs are categorically excluded from requirements to prepare National Environmental Policy Act (NEPA) documentation (i.e., Environmental Assessment or Environmental Impact Statement [see Federal Highway Administration's implementing regulations for the National Environmental Policy Act, specifically, 23 Code of Federal Regulations, 771.117(c)(9), emergency repairs under the ERFO program]).

2.6.2. Permanent Repairs

NEPA documentation (e.g., Categorical Exclusion, Environmental Assessment, Environmental Impact Statement, and Record of Decision) is the responsibility of the lead Agency administering the project.

2.6.3. Applicant as Lead or Signatory Agency

The Applicant shall request (via letter addressed to the DE) the FLHD's participation as a cooperating Agency. In the very rare case when an Applicant prepares an Environmental Impact Statement (EIS), the FLHD is required by law to be the lead or co-lead Agency. If an EIS is being considered by the Applicant, he/she shall notify the DE before issuing the Notice of Intent in the *Federal Register*.

2.6.4. FLHD as Lead or Signatory Agency

When environmental compliance is the responsibility of FLHD, Title 23, Code of Federal Regulations 771 will be the basis of the environmental process. The FLHD Project Development and Design Manual will provide additional environmental guidance. The Applicant would normally participate as a cooperating Agency at the request of the FLHD.

2.6.5. State, County or Tribe Repairs

The lead agency shall submit documentation that shows compliance with all applicable federal, state, and local environmental laws to the FLHD for approval prior to initiating construction activities funded wholly or in part by the ERFO Program.

2.7. Construction Documents

Plans, specifications, and estimates may be prepared by either the Applicant or by the FLHD. All FLHD projects shall incorporate "ERFO" and the Disaster Number in the project number, regardless of any other funding source used.

2.7.1. Repairs Not Eligible for ERFO Funding

If ineligible work is added to an ERFO project, the Applicant must:

a) Amend the DSR *and* include the source of funding in the amended document. Failure to do so may jeopardize ERFO funding.
b) Ensure that separate accounts are established.
c) Charge repairs not eligible for ERFO funding to non-ERFO accounts.
d) Report obligations/expenditures separately in the *Disaster Status Report.*

2.7.2. Simplified Plans, Specifications and Estimates Procedures

The DE or Applicant may prepare abbreviated plans and simplified project procedures to expedite ERFO repairs.

2.8. Time Limitations

ERFO projects shall be constructed promptly. 23 CFR § 668.205(e).

2.8.1. Emergency Repairs

a) Since it is in the public interest for emergency repairs to be performed as soon as possible, such repairs may be started without prior approval from the DE and shall be completed within 180 days after the event.
b) The method of contracting may be determined by the Applicant.

Special Note: Photographs must be taken prior to beginning any emergency repairs to document the extent and severity of the damage.

2.8.2. Permanent Repairs

a) *Force account* projects must be completed by the end of the second fiscal year following the event year.
b) Contracted projects must be awarded by the end of the second fiscal year following the event year. This requirement gives the Applicant 25 to 36 months to award the permanent repair construction contract and to begin construction.
c) Projects not under contract for permanent repair by the end of the second fiscal year following the event year shall be withdrawn from the POP and considered non-emergency permanent repairs unless the DE approves a time extension.

2.8.3. Contract Award Time Extensions

Time extension requests must be submitted to the FLH DE for approval at least 60 days prior to the end of the eligible period. The Applicant is responsible for submitting suitable justification for contracts that will not be awarded prior to the end of the second fiscal year following the event year. The justification must include dates of completed milestones that demonstrate the project was given prompt attention and priority over non-emergency work (see Appendix C.16).

a) The DE may grant a one year contract award time extension (see Appendix C.17) if one of the following exist:
   1) The Applicant gave the project prompt attention and priority over non-emergency work;
   2) A new disaster event significantly increases or changes an eligible site, prevents access to the site, or changes project priority; and
   3) Repairs are delayed due to ongoing litigation.
b) Contract award time extensions will NOT be granted in the following situations:
   1) Lack of available ERFO or Agency resources delay the project;
   2) Repairs proposed for Applicant's *force account* crews are delayed; and
   3) Repairs were not given prompt attention and priority over non-emergency work;

c) Time extensions beyond one year will not be considered by the DE. Beyond one year, an appeal to the FLH Associate Administrator will be required for unique or extraordinary circumstances that are beyond the control of the Applicant. Generally, the Applicant should not expect a time extension to be granted beyond the one year. This is to keep within the intent of the program and to optimize the use of available funding.

2.9. Annual Status and Disaster Closeout Reports

2.9.1. Annual Reports

The Applicant must submit two reports at the end of each year for each ERFO event.

a) The Disaster Status Report must be completed by the Applicant and sent to the FLHD office by November 30 of each year. If complete reports are not received, funding will be withheld or delayed (see Appendix C.20).
   This report serves as the Applicant's annual Status Report and final Closeout Report. When all projects have a date listed under the final column (Date Closed), the report is the Applicant's final Closeout Report.
b) An Applicant's Prior-year Unobligated Balance Report must be sent to FLH headquarters within the first quarter of the new fiscal year.
   If reports are not received or are incomplete, funding (reinstatement) of the unobligated balances will be delayed until all data inconsistencies (including prior-year obligations) have been reconciled between the Applicant the FLH program headquarters offices.

The ERFO Coordinators must summarize the results of the previous fiscal year accomplishments (e.g., Disaster Approvals and Amounts and Positive and Negative Findings, Damage Site Information, Time Extensions, Disaster Closed, Funds Obligated by Applicant, Design and Construction Workload, etc.).

The ERFO Coordinators must submit this Fiscal Year Accomplishment Report to the ERFO Program Manager by December 31 of each year.

2.9.2. Closeout

The FHWA expects that all ERFO projects will be completed by the end of the fourth year following the event year and the DE will begin closing out projects at that time. If all projects are not complete, a partial closeout of the event may be conducted.

a) *Applicant Closeout Report.* At final closeout, the Disaster Status Report must identify the actual costs of repair for each DSR. Individual sites that are combined into a single construction contract may be reported together. *Force account* work may be reported on a road-by-road basis. Combining costs from multiple DSRs, other than as stated above, is not permitted (see Appendix C.20).
b) *Partial Closeout.* At the request of the DE or the Applicant, partial closeout of completed site repairs may be initiated prior to completion of remaining site repairs.

The partial closeout may include a review of ongoing construction projects as well as completed projects.

c) *FLH Division Review.* The DE will review the Disaster Status Report for consistency with the approved POP and conduct follow-up field reviews as needed. Individual DSRs will be selected based on the following:
   1) Actual costs greater than 20 percent of the approved POP amount;
   2) High cost or technically complex projects;
   3) Force account projects; and
   4) Representative of the majority of projects.
d) Field Review Notification. The Applicant will be notified when projects are selected for an in-depth field review and shall assemble contract documents and all financial records that support the costs submitted in the Closeout Report.
   1) Contract documents include plans, specifications, permits, contract changes, contract payments, release of claims, and NEPA documentation.
   2) Financial records, such as payroll, travel, per diem; materials, supplies, equipment rental, and equipment hours must be properly documented and made available for each selected DSR.
e) Field Review Agenda. The Applicant's administration of the ERFO program will be reviewed and evaluated. The following questions will be posed to ensure compliance with applicable laws, regulations, and policies:
   1) General – What degree of support and oversight did the main (regional/state) office provide? Were projects given prompt attention and priority over non-emergency work?
   2) DSR Scope – Were projects built in substantial conformance with the approved DSRs?
   3) Completion time – Were the construction projects awarded by the second fiscal year following the event year? Were force account projects completed by the end of the second fiscal year following the event year? We're time extensions requested and approved for delayed work?
   4) DSR revisions – Were the DSRs revised when appropriate (e.g., 20 percent cost differential, change in scope)?
   5) Environmental compliance – Did the Applicant request (via letter addressed to the DE) the FLHD's participation as a cooperating Agency? Was supporting documentation that shows compliance with all applicable federal, state, and local environmental laws submitted to FLHD? How was compliance documented (e.g., CE, EA/FONSI, EIS)? If an EIS was prepared, was the FLHD notified and did the FLHD serve as the lead or co-lead Agency?
   6) Engineering – How was engineering managed (e.g., in-house, contract)? Were the design plans and specifications adequate? Did the plans and specifications have peer or Program Manager Review for compliance with the approved DSR scope? Were reviews conducted by the FLHD if noted in the approved POP?
   7) Materials and supplies – What kinds of materials and supplies were purchased? Were they eligible? Were all materials supported with adequate documentation? How were agency-furnished materials accounted for?
   8) Repairs – How were the repairs administered (e.g., in-house, contracts)? Were permanent repairs approved prior to construction? If contract modifications

occurred, were they adequately documented? Were there any salvaged items or materials?

9) Accounting
   i. Records. Is there adequate detail to determine if the charges are appropriate (e.g., location of work, who worked, time worked, and material placed or removed)? Are the agency-submitted final costs supported by their financial records?
   ii. Non-eligible work. If projects included non-eligible work, were non-ERFO costs funded and tracked separately? Were the accounts adjusted at project completion to reflect actual expenses?
   iii. Other funding sources. Were ERFO funds used to replace other funding sources (e.g., repairs programmed prior to the disaster, damage caused by legally responsible parties, share cost agreements)?
   iv. Force account rates. Are the equipment rental rates reasonable and based on an appropriate schedule?
   v. Right-of-way acquisitions. What kind of documentation is available to ensure that all requirements (e.g., Uniform Act) are met

f) FLHD Closeout Report. The DE will prepare a report summarizing the results of the office and/or field review. Adjustments to the Applicant's Closeout Report will be made if non-eligible costs were submitted. If the results of the field review indicate that additional projects should be reviewed, the Applicant will be notified and additional follow-up field reviews will be conducted as needed.

g) Closeout Letter. The DE will summarize the results of the closeout review in the Closeout Letter, revise the amount of the approved POP, and request that the ERFO Program Manager begin funding reconciliation with the Applicant's headquarters (see Appendix C.21).
   1) Submitted costs that cannot be satisfactorily assigned to a specific DSR will not be allowed.
   2) When non-eligible repairs are added to a project and the funding sources of those repairs are not satisfactorily documented, all ERFO expenditures may be disallowed.

h) Funds Reconciliation. Based on the FLHD final Disaster Closeout Letter, the ERFO Program Manager will reconcile funds with the Applicant (see Section 3.4).
   1) Funds will either be transferred to or withdrawn from the Applicant.
   2) If ERFO funds are not available, the event will not be totally closed until the funds become available and are transferred to the Applicant.

3. FUNDING: INTRODUCTION

The Highway Trust Fund (Contract Authority) is the source of Emergency Relief funds. Congress may supplement Emergency Relief funds with General funds (Budget Authority). Title 23, United States Code, Section 125, authorizes funds for emergency and permanent repairs of Federal-aid and other federally-owned roads and trails. Congress can also authorize additional emergency relief funds through supplemental appropriation acts.

The total obligation of emergency relief funds (both Emergency Relief for Federal-aid Highways and ERFO) for projects resulting from a single natural disaster or catastrophic failure is capped at $100,000,000 per event per state unless otherwise waived by Congress.

a) If the funding cap is waived or modified, the ERFO program manager will notify the Applicants and the FLHD in writing. The Applicant and the Division Engineer (DE) may then request any additional funds necessary to reach to the full amount of funding in the approved Program of Projects (POP).
b) When the Federal Highway Administration (FHWA) determines that the emergency relief cap of $100,000,000 would be exceeded if all repairs were completed, the ERFO program manager will prioritize allocation of emergency relief funds, if available, among the Applicants according to priorities in Section 3.2.3.

Applicants shall submit prioritized lists of the emergency and permanent repairs shown in the approve POP, based on available funding.

3.1. ERFO Funding Process

3.1.1. Requesting ERFO Funds

The ERFO Program Manager will only accept requests for funds from the Federal Land Management Agencies (Applicant) and Federal Lands Highway Divisions (FLHD).

Requests must be in writing (email) and must include a copy of the POP

Approval Letter (except for requests for Accelerated Release as described below).

Requests must be restricted to only those funds that will be obligated within the current fiscal year. Requested funds cannot exceed the authorized POP amount.

States, counties, and tribes cannot request funds directly from the ERFO Program Manager.

1) The BIA must request funds needed to reimburse tribes for work performed.
2) FLHD must request funds needed to reimburse states and counties for eligible work performed.

3.1.2. Types of Requests

There are three types of ERFO funds requests:

a) Accelerated Release Request. After the positive disaster finding by the FLHD and before the POP is approved, an Accelerated Release of funds may be requested. The purpose of the Accelerated Release is to make emergency repairs to re-establish vital transportation system(s) after an event. The requests will be reviewed by the FLHD on a case-by-case basis. The process to request and receive Accelerated Release is:
   1) The Applicant must make a request in writing to the FLH DE (see Appendix C.6).
   2) For each request, the Applicant will receive an Accelerated Release Response Letter. If the request is approved, the response letter will authorize the initial amount available for emergency repairs (see Appendix C.7).
   3) If Accelerated Release is approved, the Applicant's headquarter office must formally request these funds from the ERFO Program Manager.
   4) The ERFO Program Manager will transfer, if available, Accelerated Release funds to the Applicant.

b) Initial Request. Any time after the POP has been approved, the Applicant or FLHD may request the initial transfer of ERFO funds that they expect to obligate within the current fiscal year. The request shall not exceed the amount of the approved POP.

c) Annual ERFO Funds Request. This serves as a statement of upcoming needs for the next fiscal year and must be submitted by the Applicant and the FLHD to the ERFO Program Manager by September 15 (see Appendix C.18).
   (1) The ERFO Program Manager will consolidate these requests and submit them to the Emergency Relief (ER) Program Manager.

The ERFO Program Manager will transfer to the Applicant or allocate to the FLHD offices amounts not to exceed the approved POP.

3.1.3. Allocation of ERFO Funds/Obligation Authority

All ERFO fund allocations shall be entered into the Department of Transportation (DOT) accounting system by state and funding program code. The ERFO Program Manager will allocate Obligation Authority, by disaster, state, and funding program code to the Applicant's headquarters or the FLHD as described below.

Special Note: Allocation of ERFO funds will be delayed if prior year Disaster Status Reports have NOT been submitted to the FLHD. In addition, allocation of ERFO funds may also be delayed until prior-year obligations have been reconciled between the Applicant and the FLH program headquarters offices.

a) Applicants will receive an Allocation of ERFO Funds letter with a summary of the Applicant's current fiscal year ERFO program obligations. In addition, the FHWA Office of Budget and Finance shall issue a Form 370 (Allotment Advice Form) to the Applicant, confirming the allocation authority and limitation.

b) FLHD will receive an Allocation Memorandum from the ERFO Program Manager with a summary of the FLHD's current fiscal year ERFO program or obligations.

c) Each ERFO Allocation Memorandum will provide funding program codes that indicate the type of ERFO funds authorized. The FLHD and the Applicant must use these funding program codes to track the obligation and expenditure of all ERFO funds received. The Applicant and FLHD must report the obligation of funds to the ERFO Program Manager by event number and funding program code(s). The three-digit program fund code historically used follows the format PC-09[1] and TR6[1].

3.1.4. Fund Transfers

a) State or local governments will enter into a Reimbursable Agreement with the FLHD. With this method, ERFO funds needed for the repair costs are allocated by an Allocation Memorandum to the FLHD by event number, state, and funding program code. Work is done on a reimbursable basis and paid per provisions in the agreement.
b) Indian Tribal Governments will enter into a PL 93-638 contract with the Bureau of Indian Affairs (BIA) for repairs performed by the tribe. The BIA has the responsibility of an Applicant for Tribal-administered projects and will request and receive funds from the ERFO Program Manager for Tribal projects.

3.2. Advancing Projects during an ERFO Fund Shortage

When all available funds are exhausted, ERFO fund requests will be held until the U.S. Congress authorizes additional funds. Recognizing that prompt congressional action is not always possible, the following options or combinations of options may be used to fund or advance ERFO projects on an interim basis.

3.2.1. Unobligated Balances from other Approved Disasters

If an Applicant has received funds for an ERFO disaster, the unobligated balance can be used on other approved disasters. The Applicant is required to obtain approval from the ERFO Program Manager when this option is needed. This use of ERFO funds is not allowable when funds are limited by law to a specific disaster or the use of the funds exceeds the approved POP for that disaster.

3.2.2. Other Funds

a) *Applicant funds.* Other Applicant funds may be used to perform approved ERFO repairs with reimbursement amount equal to the total obligated amount of Applicant funds.
b) *Federal Lands Highway Program (FLHP) Funds.* FLHP funds may only be used on highways that are eligible for that particular category of funds. In addition, the following approval is required:

    1) Forest Highway (FH) funds may only be used upon prior approval of the tri-agency agreement from the appropriate state, Forest Service, and FLHD officials;
    2) Park Road and Parkways (PRP) funds may only be used upon prior approval from National Park Service Headquarters;

3) Indian Reservation Roads (IRR) funds may only be used upon prior approval from the affected Indian tribal government(s) and the BIA;
4) Refuge Roads (RR) funds may only be used upon prior approval from the U.S. Fish and Wildlife Service Headquarters office; and
5) Public Lands Highway Discretionary (PLH-D) funds are project-specific and approval from the FHWA is required.

3.2.3. ERFO Funding Priorities

When available ERFO funds are insufficient to meet current fiscal year fund requests, the ERFO Program Manager will allocate funds as shown below. Listed in priority order, they are:

a) Reimbursement (payback) of Applicant funds, including state, county, local and tribal government funds used on approved ERFO repairs;
b) Reimbursement of FLHP funds (FH, PRP, IRR, RR, and PLH-D) used on approved ERFO repairs;
c) Construction engineering and contract modifications for ongoing ERFO construction projects;
d) Emergency repairs of high-use facilities that are critical to the agencies' mission and/or health and welfare of the local community/region; and
e) Fiscal year disasters.

3.3. Fiscal Responsibility and Reporting Requirements

The Applicant is responsible for the administration of ERFO funds and the obligation authority that has been transferred with it.

This includes compliance with all applicable laws, regulations, and fiscal reporting requirements.

3.3.1. Projects with Combined Funding

When a construction contract includes both eligible and ineligible ERFO work, the Agency administering the contract has the following responsibilities.

a) Applicant, state, county, and tribe administered projects. The Agency administering the contract must inform the FLHD of the funding source of the ineligible work in writing prior to project completion. Failure to do so may cause the entire site to become ineligible for ERFO funding. The responsible Agency must ensure that separate accounts are established and repair not eligible for ERFO funding is charged to non-ERFO accounts.
b) FLHD Administered Projects. The Agency responsible for the operation and maintenance of the road must transfer funds (obligation authority and liquidated cash) or enter into a reimbursable agreement with the FLHD for the ineligible work prior to the FLHD advertising the project. Fund transfers will be made via Form-1151 or other established fund transfer procedures.

3.3.2. Reporting Requirements

Each event requires a Disaster Status Report to be submitted to the ERFO Coordinator documenting the status of all approved funds.

Special Note: The ERFO Program Manager will withhold funds if the reporting requirements shown below are not met.

a) *Unobligated Balances.* The Applicant and FLHD must report all unobligated ERFO funds at the end of each fiscal year as follows:
   1) *Applicant.* The Applicant shall report the amount of unobligated ERFO funds (by program code) to the FHWA using a Standard Form 133, Report on Budget Execution (generally done by the Applicant's Budget Office by September 30). The ERFO Program Manager and the Applicant will reconcile all reported amounts. Per written request by the Applicant, the ERFO Program Manager will reinstate the unobligated fund balances (by event number and program code) as soon as possible following the reconciliation process.
   2) *FLHD.* The ERFO Program Manager and the FLHD will reconcile all reported amounts of unobligated ERFO funds (by event number and program code). Per written request from either the FLH DE or the ERFO Coordinator, the ERFO Program Manager will withdraw all unobligated funds and reinstate the funds (by event number and program code) as soon as possible following the reconciliation process.

b) *Disaster Status Report.* The Applicant and the FLHD must prepare and submit a Disaster Status Report for each disaster by November 30 of each year. The information on this report should reflect the status of each disaster through the end of the previous fiscal year.
   1) Applicant. Annually submit the Disaster Status Report to the FLHD. As a final closeout report by the Applicant, the Disaster Status Report must present a final accounting of the costs claimed for the event by each DSR. The final Disaster Status Report must be submitted to the ERFO Program Coordinator within six months of the completion of all repairs (see Appendix C.20). Supplemental funds used to address agency priorities not approved for ERFO funding must also be reported. The BIA must include the status of work performed by tribes.
   2) FLHD. Submit the Disaster Status Report for any work conducted by the FLHD to the Applicant. In addition, submit these and all reports received from the Applicants to the ERFO Program Manager. As a final closeout report by the FLHD, a Disaster Status Report must present a final accounting of the cost claimed for the event by each DSR. The final closeout report must be submitted to the Applicant and the ERFO Program Manager within six months of the completion of all repairs (see Appendix C.20).

c) *Closeout Supporting Documentation.* Costs that cannot be satisfactorily assigned to a specific DSR will not be reimbursed. For each DSR, provide the following:
   1) *Labor.* Time keeping document (time slip, time and attendance report, etc.);
   2) *Material.* Stock slip, vendor invoice;
   3) *Equipment.* Mobilization, operation costs, depreciation or replacement costs, and rental costs;

4) *Contracts.* Plans, specifications, payments, construction modifications, and copy of executed contracts; and
5) *Expenditures.* Print out from Agency's financial system.

3.4. Fund Reconciliation

Based on the FLHD final Disaster Closeout Letter, the ERFO Program Manager will reconcile funds with the Applicant's headquarters and the FLHD.

a) *Funds Available.* Funds will either be transferred to or withdrawn from the Applicant or allocated to or de-allocated from the FLHD by event number, state, and program code.
b) *Funds not available.* If ERFO funds are not available for transfer or allocation, the event will not be closed until funds become available and are either transferred to the Applicant or allocated to the FLHD by event number, state, and program code.

4. Eligibility: Introduction

The intent of the ERFO program is to fund repair of *serious* damage to roadways caused by a natural disaster or catastrophic failure. ERFO funds will not be approved to supplant other funds for the correction of pre-existing, non-disaster related deficiencies nor will ERFO funds be approved for the repair or reconstruction of facilities affected by long term, pre-existing conditions, or predictable developing situations. Damage will be approved on a site-by-site basis and will not include work considered to be heavy maintenance. Betterments added to prevent future damage must be economically justified.

Repair costs of a damaged site (emergency and permanent repairs) totaling less than $5,000 are NOT eligible for ERFO funding. In addition, *damages at two or more sites shall NOT be combined to meet that threshold.*

4.1. Federal Roads and Trails

Roads and those trails that are high use engineered surfaced; Americans with Disabilities Act - ADA compliant; listed on the National Trails System Act of 1968 (see Appendix E) and defined in this section are eligible for funding.

4.1.1. Eligible Roads and Trails

a) *Bureau of Land Management* roads and trails that are classified by the Bureau of Land Management as "Land Management Roads," as well as those roads and trails on the revested Oregon and California Railroad and reconvened Coos Bay Wagon Road Grant Lands in Oregon (commonly called O and C Lands) are considered public lands development roads and trails (as per memorandum of agreement).
b) *Bureau of Reclamation* public roads owned by the Bureau of Reclamation on its project lands meet the definition of Public Lands Highways (as per memorandum of Agreement).
c) *Corps of Engineers* public roads owned by Corps on water resources development projects and other administrative units meet the definition of Public Lands Highways (as per memorandum of Agreement).
d) *Forest development roads and trails.* - The terms "forest development roads and trails" means forest roads and trails under the jurisdiction of the Forest Service.

e) *Forest highway.* - The term "forest highway" means a forest road under the jurisdiction, and maintained by, a public authority and open to public travel.

f) *Forest road or trail.*- The term "forest road or trail" means a road or a trail wholly or partly within, or adjacent to, and serving the National Forest System that is necessary for the protection, administration, and utilization of the National Forest and the use and development of its resources.

g) *Indian reservation road.*- The term "Indian reservation road" means a public road that is located within or provides access to an Indian reservation or Indian trust land or restricted Indian land that is not subject to fee title alienation without the approval of the Federal Government, or Indian and Alaska Native villages, groups, or communities in which Indian and Alaskan Natives reside, whom the Secretary of Interior has determined are eligible for services generally available to Indians under Federal laws specifically applicable to Indians and is on the Bureau of Indian Affairs' Indian Reservation Road Inventory.

h) *Military installation.*- The term "military installation" means public roads owned by the United States DOD military services on military installations meet the definition of Public Lands Highways (as per memorandum of Agreement).

i) *Park road.* – The term "park road" means a public road, including a bridge built primarily for pedestrian use, but with capacity for use by emergency vehicles, that is located within, or provides access to, an area in the National Park System with title and maintenance responsibilities vested in the United States.

j) *Parkway.* – The term "parkway" means a parkway authorized by Act of Congress on lands to which title is vested in the United States.

k) *Public lands development roads and trails.* – The term "public lands development roads and trails" means those roads and trails that the Secretary of the Interior determines are of primary importance for the development, protection, administration, and utilization of public lands and resources under the control of the Secretary of the Interior.

l) *Public lands highway.* – The term "public lands highway" means a forest road under the jurisdiction of and maintained by a public authority and opens to public travel or any highway through un-appropriated or unreserved public lands, nontaxable Indian lands, or other Federal reservations under the jurisdiction of and maintained by a public authority and open to public travel.

m) *Refuge road.* - The term "refuge road" means public roads owned by the U.S. Fish and Wildlife Service on national wildlife refuges, national fish hatcheries, and other administrative units such as the National Conservation Training Center, meet the definition of Public Lands Highways (as per memorandum of Agreement).

n) *Western Area Power Administration roads.*- The term " Western Area Power Administration roads" are public roads which provide access to WESTERN-owned facilities such as transmission lines, switch yards, and substations meet the definition of public lands highways (as per memorandum of agreement).

4.2. Closed Roads

Roads that have been closed and/or placed into storage will not be approved for ERFO funds.

4.3. Repair Standards

Roads and trails will generally be repaired to restore safe passage consistent with the pre-disaster conditions of the facility.

4.4. Road Damage

4.4.1. Boat Ramps

Boat ramps are not eligible for ERFO funding, but the roads leading to the ramps may be eligible.

4.4.2. Bridges

a) General:
   1) When a bridge is destroyed or repairs are not feasible, the cost of a comparable facility that meets current geometric and construction standards for the type and volume of traffic it will carry during its design life is eligible.
   2) Bridges that are scheduled for replacement (included in the 4-year STIP) are eligible for reasonable Emergency Repairs to restore essential traffic, but are not eligible for Permanent repairs.
   3) Only repair of disaster related damage is eligible. If repair of disaster damage would not correct pre-disaster structural deficiencies, then Permanent Repairs are only eligible if other funds are used to simultaneously correct the structural deficiencies.

b) Temporary/Prefabricated/Portable bridges will be approved based on the following conditions:
   1) Placement of the temporary bridge is needed to maintain essential traffic and failure to place the bridge would result in a disastrous impact.
   2) Bridge may be leased or purchased contingent upon the following stipulations:
   i) Purchase price minus resale value only.
   ii) An economic analysis is completed prior to the purchase that demonstrates significant cost benefit of purchasing over leasing.
   iii) Removal of the structure following permanent repairs.

Special Note: Maintenance of bridges after installation is not eligible.

4.4.3. Culverts

Repairing, cleaning, or resetting damaged culverts is eligible for ERFO funds. Destroyed culverts may be replaced to current engineering design standards or environmental requirements. For aquatic organism passage, see Section 4.5.14.2 Environmental Requirements.

4.4.4. Embankments

Embankment replacement to repair serious damage is eligible under the following conditions:

a) Failures extending into the traveled way.
b) Failures NOT extending to the traveled way are eligible if the roadway is in imminent danger of failure or if the damage severely impairs safety or usefulness of the roadway.

4.4.5. Levees and Dikes

Levees and dikes used for water control are not eligible. Roads on top of these structures are eligible (usually limited to the top 12 inches) after repair to levees or dikes are completed.

4.4.6. Parking Lots

Parking lots at a road terminus, that have an engineered surface, are eligible for ERFO funding.

4.4.7. Roadway Surfacing

Roadway surfacing, if major loss has occurred, is eligible. Roadway surface upgrades beyond the pre-disaster conditions are not eligible for ERFO funds.

ERFO funds are not intended for repairs of surface damage caused by traffic whether or not the damage was aggravated by saturated sub-grade or inundation.

4.4.8. Stone Masonry

a) Stone Walls (Substandard). A replacement wall meeting current standards is eligible for ERFO funds if the applicant agrees to reconstruct the entire wall to current standards and uses non-ERFO funds to replace the undamaged portion(s) of the wall. However, if small section were damaged, ERFO funds could be used to repair the wall to pre-disaster conditions.
b) Stone Masonry repairs on walls meeting current standards are eligible provided the applicant is able to document the pre-disaster condition of the masonry and the stone masonry damage is the result of the event.
c) Stone masonry on new features (walls, curbs, etc.) is only eligible if the damaged/destroyed feature had stone masonry.

4.4.9. Basin Flooding

According to 23 CFR 668.109(b)(8), raising the grade as a result of basin flooding may be eligible for ERFO funds if the FHWA makes a formal determination based on the following factors:

a) The affected road must be considered a critical Federal-aid highway or transportation facility with potential long-term loss of use due to an unprecedented rise in basin water level and if the corrective work is restorative rather than preventive in nature.
b) The proposed increase in grade eligibility must be those routes that are critical to restoring traffic service. Factors to take into account in evaluating whether individual routes are critical could include provision of essential community services such as access for school, ambulance, fire, and mail vehicles, availability of alternate routes, length of detours, etc.

4.4.10. Slides

Eligible repairs include removing debris, restoring slopes, and reconstructing the roadway and drainage facilities to provide a safe usable roadway. Slides that occur after the disaster period may be eligible for ERFO funds if written documentation shows that the slides were the direct result of the disaster event.

Existing slides are eligible only if the slide is significantly accelerated by the disaster. Pre-existing, slow moving slides that are not significantly accelerated during the event are not eligible.

Slide stabilization is betterment.

a) ERFO funds may be approved for slide stabilization based on a detailed analysis of the slide, by the applicant. ERFO funds may not be provided for all corrective measures performed by the applicant to "stabilize" the slide. ERFO funds generally are limited to paying the cost to provide a usable road.
b) The analysis by the applicant must show that the stabilization cost does not exceed anticipated future ERFO costs to remove debris and provide a usable road. The cost of the analysis is the responsibility of the Applicant agency.
c) The cost of monitoring slide stabilization measures after construction is NOT eligible for ERFO funds.
d) A recurring slide previously stabilized with ERFO funds will not be considered eligible for additional funding, except when expressly approved by the DE.

4.5. Subsidiary Costs

4.5.1. Additional Damage

Additional damage to a previously-approved site may be eligible.

a) Damage that occurs before permanent repairs are completed is eligible for ERFO funds provided:
    1) The additional damage could not have been anticipated and appropriate emergency ERFO repairs were made to prevent further damage.
    2) The repairs are not the responsibility of the contractor.
b) When additional damage is the result of a subsequent eligible disaster event, the repairs are eligible. The repairs may be combined under either event as follows:
    1) If the cost of the additional damage is applied to the original disaster, an amended Damage Survey Report (DSR) is required.
    2) If the cost of the additional damage is applied to the new event, a new DSR is required and the original DSR for that site must be amended.
c) Damage that occurs after permanent repairs are completed may be eligible if it can be shown that the damage was directly caused by the event and not by other factors, and the repair shows signs of failure within three months after final inspection.

4.5.2. Administrative Expenses

a) Permanent and any temporary Employees. Regular salaries and overtime salaries and wages of all permanent and temporary employees directly engaged in the performance of work on ERFO projects are eligible for reimbursement. Timekeeping procedures must allocate employees' time to a specific site, each day on an hourly basis. A timekeeping document, such as time slip, time and attendance report, or time book, must be available for examination by audit personnel to support direct labor costs claimed. A responsible employee having knowledge that the time distribution is accurately reported should sign the document.

b) Payroll Additives. Usually referred to as labor surcharge, a set percentage over and above the total direct labor costs charged to a project is eligible for ERFO participation. This surcharge covers costs of various types of leave allowances, industrial accident insurance coverage, and other employee benefits. The allowable percentage rates will normally vary from year to year. Also, employee benefits allowed by one applicant may differ widely from those allowed by another. Therefore, the records used in developing percentage rates should be preserved under suitable control conditions to percentage rate may be applied only to direct labor costs.
c) Travel and Per Diem expenses properly assigned to an eligible site and directly attributable to individual ERFO projects in accordance with Federal Joint Travel Regulation – are eligible for ERFO funding.

4.5.3. Aesthetic Treatments

Aesthetic treatments on new walls, bridges, and roadways such as stone facing or stone curb are not eligible.

4.5.4. Betterments

Betterments (added features) that help protect highway facilities from possible future ERFO eligible damage may be eligible if the applicant demonstrates through a life cycle cost analysis that it is economically justified to the ERFO program.

a) Requests for funding betterments must be submitted by the applicant to the FLH DE for approval. A justification and economic analysis should be provided at the time the Damage Site Survey Team prepares the DSR.
b) Life Cycle Cost Analysis. Betterments must be clearly justified to the ERFO program through a life cycle cost analysis. The analysis must be based solely on the present and future expected cost to the ERFO program. The analysis cannot include other factors typically included in highway benefit/cost evaluations, such as maintenance costs, traffic delay costs, added user costs, motorist safety, and economic impacts. The analysis period will vary depending on the significance of the facility. Bridges and roads will typically be 50 years and 25 years, respectively, with some roads as low as 10 years or less.

Special Note: In many instances betterments will fail to meet the test of being economically justified to the ERFO program. If betterment provides considerable benefit when other factors are considered, the applicant is encouraged to use other funds, as appropriate, to fund it.

These other funds must be identified to the FLH DE on status reports.

c) Examples of Betterments. The following examples may help protect highway facilities from possible future damage and as a result may be approved:
   1) Raising roadway grades.
   2) Relocating roadways to higher ground or away from slide-prone areas.

3) Stabilizing slide areas.
4) Stabilizing slopes.
5) Installing riprap.
6) Lengthening or raising bridges to increase waterway openings.
7) Deepening channels.
8) Increasing the size or number of drainage structures.
9) Replacing culverts with bridges.
10) Installing seismic retrofits on bridges.
11) Adding scour protection at bridges.
12) Adding spur dikes.

4.5.5. Clearing Debris/Timber Removal

Clearing debris and timber from roadways and water courses around bridge piers to adjacent locations within the right-of-way are eligible for ERFO funds. The costs of hauling timber to sawmill or to other locations beyond the proximity of the highway right-of-way are not eligible for ERFO funds.

a) Debris removed on roads is limited to that necessary to assure the safety and full function of the roadway.
b) Debris removed from watercourses is limited to that necessary to protect the structural integrity of the road or bridge and is limited to debris within the actual drainage channel, right-of-way boundary (or typical boundary if none).

4.5.6. Construction Road Damage

Access road (e.g., detour roads, haul roads, etc.) damage is eligible provided the applicant is able to document the pre-disaster condition of the roadway, legal load limits were enforced, and the damage is the result of the following:

a) Damage caused by vehicles hauling roadway construction equipment or material necessary for making emergency ERFO repairs.
b) Damage caused by emergency vehicles responding to the event.
c) Damage to the roadway caused by other traffic is NOT eligible.

4.5.7. Damage beyond Highway Right-of-Way

Damage beyond the normal highway right-of-way width, including damage that is caused by or attributed to the highway facility (even if the applicant may be found liable for the damage) is not eligible.

4.5.8. Decommissioning or Storing

Decommissioning or storing eligible roads in lieu of repair is eligible. Funds are capped at the cost to repair eligible damage on the road. A portion of the decommissioned road can be left to function as a trail; however, trail construction along the abandoned facility is NOT eligible. Suitable turnaround facilities at the beginning of the decommissioned road section are eligible. Re-contouring is not eligible except at stream crossings.

Roads identified in a Decommissioning Plan will NOT be approved for ERFO funding. Decommissioning costs should be funded as originally planned.

4.5.9. Delayed Projects

Costs expended after the eligible period are not eligible for funding unless a time extension has been granted.

4.5.10. Detour Routes

a) Construction costs are eligible to restore essential traffic consistent with anticipated volume and type of traffic, considering appropriate speeds, safety considerations, duration of use, and overall costs. However, detour construction that establishes pre-disaster levels of service does require prior approval by the FLH DE.
b) Routine Maintenance of detour routes, not related to increase traffic volumes such as mowing, brushing, maintaining drainage, pavement signing and snow and ice removal are not eligible for ERFO funding.

4.5.11. Direct Costs

Direct costs incurred to repair damage are eligible. These costs include salaries, wages, fringe benefits, consultant services, travel, materials, supplies, and equipment directly used to accomplish the repair of eligible damage.

All costs must be properly documented in accordance with acceptable accounting standards (e.g., time slips, time and attendance reports, time books, contracts, receipts) and readily available for inspection by the ERFO coordinator. For additional details, see Section 4.5.2.

4.5.12. Emergency Repairs

Emergency repairs are eligible and shall reduce to the greatest extent feasible, the cost of permanent repairs.

4.5.13. Engineering Cost

Engineering costs directly attributed to repairs are eligible as follows:

a) Preliminary engineering.
b) Construction engineering related to contract administration and inspection of construction work.

4.5.14. Environmental Requirements

a) General

When project scope changes are required to avoid, minimize, or mitigate impacts related to the repair or reconstruction of eligible roads, the additional costs are fully funded up to 100% of the least cost environmentally viable alternative to current standards when fully documented.

b) Aquatic Organism Passage
   1) Cost for upsizing culverts or making other modifications to accommodate aquatic organism passages is eligible; and
   2) Cost of major improvements (e.g., replacing a culvert with a bridge), will not be considered eligible for ERFO funding, except when expressly approved by the DE.

c) Documentation

The Applicant must submit the following:

1) The specific section of the law, regulation, policy, or treaty that prohibits repair in-kind;
2) The specific impacts of the repair in-kind alternative that must be avoided, minimized, or mitigated;
3) A project description and cost estimate of the least cost, environmentally viable project that will meet the minimum requirements;
4) FLMA certification that the project is in the public's best interest; and
5) Results of consultation with regulatory agencies, if applicable.

4.5.15. Equipment, Materials and Supplies

a) Equipment costs may be eligible as follows:
   1) Purchase (minus resale value) must be approved in advance by the FLH DE and must be cost effective when compared to renting.
   2) Rentals may include applicant-owned equipment, equipment rented from another public entity, and equipment rented from commercial sources. Rental rates should include operation, maintenance, and depreciation costs. Documentation required to support costs include the following:
      i) A commercial invoice or other applicant-generated time and equipment use report that specifies, for each DSR the equipment was used on, the date of use and hours of operation.
      ii) The equipment use record must be signed by the applicant's employee.

b) Office materials and supplies are eligible for ERFO funds provided they are:
   1) Expendable.
   2) Properly identified on the stock-issue slip or vendor's invoice with the project charged and readily available for inspection by the FLHD ERFO coordinator.

c) Construction equipment and materials that are NOT eligible for ERFO funds include:
   1) Replenishing or paying losses for damaged or destroyed material stockpiles (i.e., sand gravel, pipe, etc.) and items awaiting installation not incorporated into the roadway.
   2) Damages to an applicant's equipment (i.e., motor graders, tractors, etc.) and other losses (i.e., maintenance sheds, offices, picnic tables, trash cans, tent pads, boat launch ramps, etc.).

4.5.16. Force Account Repair Costs

Force account repair costs that exceed the costs comparable to those expected by the competitive bidding process are not eligible.

4.5.17. Heavy Maintenance

Work usually done to repair damage normally expected from seasonal and occasionally unusual natural conditions or occurrences is considered heavy maintenance.

This includes work at a site required as a direct result of a disaster which can reasonably be accommodated by a state or local road authority's maintenance, emergency or contingency program even though it exceeds the minimum site threshold of $5,000.

4.5.18. Mobilization

Mobilization costs are eligible as calculated from:

a) The pre-disaster location of the equipment to the first eligible site.
b) From one eligible site to the next logical eligible site.

4.5.19. Permanent Repairs

Permanent repairs performed prior to authorization are not eligible, unless it is determined to be more economical or practicable to perform such work as an associated part of the emergency repairs. Permanent repairs that differ from the approved repairs are not eligible.

4.5.20. Programmed Work

Programmed work to repair or replace facilities damaged during a disaster is not eligible.

This work must be financed as originally intended.A project is considered "programmed" if the construction phase is included in the FLHD program or State Transportation Improvement Program or if contract plans are being prepared.

4.5.21. Preventative Work

Preventative work to avoid damages is not eligible unless approved as betterment.

4.5.22. Right-of-Way

Right-of-way acquisition required for repair of eligible damage is eligible for ERFO funding.

4.5.23. Stream Channel Reconstruction

Restoration of the stream channel necessary for the continued safe operation of bridge and drainage facilities is eligible for ERFO funding.

4.5.24. Studies

Studies undertaken to make transportation or land management decisions are not eligible. These include studying and gathering information for:

a) Access Travel Management.
b) Alternatives not approved for ERFO funding.
c) Long-term environmental impacts such as climate change.
d) Road closure/decommissioning.

4.5.25. Traffic Control

Traffic control devices, barricade erection and removal, and flagging traffic through or around damaged section of roadway or bridges are eligible.

4.6. Replacement Facilities

Replacement facilities are appropriate when it is not practical and economically feasible from an engineering perspective to restore a damaged element to its pre-existing condition.

4.6.1. Design Standards

Replacement facilities may be built to current standards as described below. (Reference 23 CFR § 668.209(d) Eligibility of work.).

a) Roads - current engineering design standards of comparable capacity and character to the destroyed facility.
b) Bridges - current design standards of a facility that meets current geometric and construction standards for the type and volume of traffic it will carry during its design life.
c) Walls - current engineering design and construction standards and does not include cosmetic features such as stone facing.

4.6.2. Funding

ERFO funds are limited to the least cost alternative that meets the design standards. The Applicant may supplement ERFO funds with other funds to construct an alternate facility.

4.7. Resolving Eligibility Issues

Resolving eligibility issues shall be handled between the FLHD and applicant ERFO coordinators.

4.7.1. Unresolved Disputes

Unresolved disputes may be forwarded to the DE.

4.7.2. Appeals

Appeals may be elevated to the FLH Associate Administrator with copies to the following:

a) Director, Office of Program Development
b) ERFO Program Manager

5. Roles and Responsibilities

5.1. Applicant

The applicant's representative is responsible for coordinating ERFO activities at the following organizational levels:

a) Field
   1) Takes photos of the damage prior to repairs to help justify and document site eligibility.
   2) Initiates emergency repairs.
   3) Advises higher authorities about the severity and extent of the damage.
   4) Participates on disaster assessment teams and provides all available information required for the Disaster Assessment Report.
   5) Arranges for advance briefing sessions.
   6) Ensures that all teams evaluate the damage in a consistent manner.
   7) Monitors the progress of Damage Survey operations.
   8) Completes all required documentation from the beginning to end for the Program of Projects (see Section 2).
   9) Requests approval to use force account.
   10) Oversees the development of Plans, Specifications, and Estimates.
   11) Designates the Contracting Officer for work being administered by the applicant.
   12) Concurs in award/rejection of bids.
   13) Accepts completed construction.
   14) Insures prompt and efficient completion of ERFO repairs.
   15) Requests funds from Regional/State offices.
   16) Administers transfer of ERFO funds from applicant's headquarters office.

17) Recovers ERFO repair costs from legally responsible parties and contacts the ERFO Program Manager about re-payment policies.
18) Submits the following to the Federal Lands Highway Division office through the applicant's Regional/State office:
    i) Notice of Intent Letter.
    ii) Program of Projects.
    iii) Requests for changes.
    iv) Plans, Specifications, & Estimates.
    v) Suitable justification for all requests for Time Extension(s).
    vi) Disaster Status Report.
    vii) Any request for Federal Lands Highway assistance in project development and/or construction process.

b) Regional/State Office is responsible for the following activities:
   1) Notifies the Federal Lands Highway Division's ERFO Coordinator of a natural disaster or catastrophic failure.
   2) Arranges Disaster Assessment activities.
   3) Assignment of personnel to the Damage Survey Team.
   4) Participates in Disaster Assessment activities in the field.
   5) Assembles the following:
      i) Program of Projects.
      ii) Revisions for the disaster.
   6) Prepares all ERFO correspondence for the applicant field unit including but not limited to the following:
      i) Notice of Intent Letter.
      ii) Program of Projects Letter.
      iii) Time Extension requests.
      iv) All Disaster Status Reports (including partial).
      v) Any request for Federal Lands Highway assistance in project development and/or construction process.
   7) Monitors and ensures prompt and efficient completion of all repairs.
   8) Provides:
      i) Technical assistance.
      ii) Copies of this ERFO Manual.
      iii) Training.
      iv) Information.
      v) Eligibility guidance to applicant (field units).
      vi) Maps identifying public and administrative routes.
   9) Requests funds through the applicant's headquarters office.

c) Headquarters
   1) Advises applicant field units of the amount of ERFO funds available for their disasters.
   2) Requests the following from the Federal Lands Highway Headquarters office:
      i) ERFO fund transfers.
      ii) Liquidating cash.
   3) Reinstatement of year-end unobligated ERFO fund balances.

4) Reports annual obligations and expenditures of ERFO funds to Federal Lands Highway Headquarters office.
5) Concur that roads maintained for high clearance vehicles, trails, and public lands development administrative roads (on lands under the control of the Secretary of Interior) are a national priority.

5.2. Federal Lands Highway

a) Division ERFO Coordinators are primarily responsible for coordination of the ERFO Program and subsequent activities with the applicant's Regional/State office. They are also responsible for ERFO program issues and activities within the Federal Lands Highway Division office and with the Federal Lands Highway Headquarters office including but not limited to the following:
   1) Receives informal Notices of Intent from the applicant.
   2) Contacts other Federal Agencies to see if they have related damage.
   3) Notifies the Federal Lands Highway Division Engineer.
   4) Notifies Federal Lands Highway ERFO Program Manager.
   5) Writes the Notice of Intent, Acknowledgment Letter.
   6) Conducts Disaster Assessments to verify the significant impact of the natural disaster or catastrophic failure and then coordinates all Disaster Assessment activities necessary for preparing Disaster Assessment Reports.
   7) Writes the Disaster Assessment Report.
   8) Provides joint guidance briefings prior to beginning the Damage Surveys.
   9) Coordinates Damage Survey Team(s) assignments.
   10) Instructs the Damage Survey Team(s) about the following:
      i) How to perform the detailed damage site inspections.
      ii) How to prepare the Damage Survey Reports.
      iii) Eligibility requirements.
   11) Reviews and ensures completeness of all pertinent documentation including, but not limited to, the following:
      i) Damage Survey Reports.
      ii) Applicant Program of Projects and subsequent revisions.
      iii) Disaster Status Reports.
      iv) Time Extension requests.
   12) Makes recommendations to the Federal Lands Highway Division Engineer.
   13) Discusses eligibility questions with the Federal Lands Highway Division Engineer.
   14) Forwards eligibility questions that cannot be resolved between the Federal Lands Highway Division Engineer and the applicant to the Federal Lands Highway ERFO Program Manager for resolution.
   15) Spends as much time in the field as practical when Damage Survey Reports are being prepared.
   16) Retains the original of each Damage Survey Report and furnishes a copy of each completed Damage Survey Report to the applicant.
   17) Reviews the applicant's Program of Projects.

18) Writes and forwards the Program of Projects Approval Letter and corresponding spreadsheets to the Federal Lands Highway Division Engineer for signature/ approval.
19) Writes ERFO correspondence for the Federal Lands Highway Division Engineer.
20) Reviews and recommends approval of a variety of documents including, but not limited to, the following:
    i) Applicant's Disaster Status/ Reports.
    ii) Completed ERFO projects.
    iii) Plans, Specifications, and Estimates on selected projects.
    iv) Damage Survey Reports and their amendments.
    v) Program of Projects and their revisions.
    vi) Time Extensions.
21) Monitors in-house design efforts.
22) Requests funds from Federal Lands Highway Headquarters office.
23) Administers allocated ERFO funds.
24) Serves as Contracting Officer's Technical Representative (COTR) on Reimbursable Agreements.
25) Prepares Disaster Status Reports for in-house work.
26) Forwards Disaster Status Reports to the Federal Lands Highway ERFO Program Manager.
27) Conducts Disaster Closeout reviews in a timely manner, makes closeout recommendations to the Federal Lands Highway Division Engineer, and prepare final Closeout Letters.
28) Prepares and Forwards Fiscal Year Accomplishment Report to the Federal Lands Highway ERFO Program Manager.

b) Division Engineer - Is ultimately responsible for successful administration of the ERFO Program within the Federal Lands Highway Division office. Specific duties include, but are not limited to, the following:
1) Designates the Federal Lands Highway Division's ERFO Coordinator.
2) Acknowledges the applicant's Notice of Intent Letter.
3) Designates and dispatches Disaster Assessment Team(s), as necessary.
4) Issues Positive or Negative Findings.
5) Determines ERFO eligibility.
6) Approves Damage Survey Reports, Program of Projects and subsequent changes/amendments.
7) Delivers the ERFO Program for in-house work.
8) Approves the following:
    i) Requests for Time Extensions.
    ii) Plans, Specifications, and Estimate package.
    iii) Force Account work.
9) Concurs in award/rejection of bids.
10) Accepts completed construction for repairs performed by the Federal Lands Highway Division.
11) Requests funds from Federal Lands Highway Headquarters office.
12) Administers allocated ERFO funds.
13) Completes disaster closeout in a timely manner.

14) Enters into reimbursable agreements with agencies, State, county and local governments, prior to performing work.
15) Ensures compliance with NEPA and all applicable environmental law and regulations.
16) Forwards final Disaster Repair Closeout Report and letter to the Federal Lands Highway office.

c) ERFO Program Manager
1) Administers the nationwide ERFO Program and assures overall program quality through management oversight, program stewardship and funds accountability.
2) Coordinates with the Federal Highway Administration's Emergency Relief (ER) Program Manager the following :
    i) Annual ERFO fund allocations and transfers.
    ii) Disaster Numbers.
    iii) Reporting requirements.
    iv) ERFO Funds Request for fiscal year.
    v) Resolves conflicts between the ER and ERFO Programs.
3) Reviews Positive or Negative Findings.
4) Transfers and allocates funds.
5) Coordinates with applicant, division, and headquarters personnel.
6) Issues policy and procedures.
7) Reviews eligibility appeals from applicants and makes recommendations to the Director, Office of Program Development and the Associate Administrator, Office of Federal Lands Highway.
8) Updates the ERFO Manual.
9) Develops, conducts, and/or assists in ERFO training.
10) Develops and monitors Interagency Agreements.
11) Assists applicants in the reconciliation and reinstatement of end-of-year unobligated ERFO funds by Program Code.
12) Coordinates with the applicant's headquarters offices and the Federal Lands Highway Division's ERFO Coordinators on monitoring program activities.
13) Develops:
    i) Method to track fund transfers and allocations.
    ii) Fiscal year obligation projections.
14) Leads the following efforts:
    i) Overall ERFO quality assurance.
    ii) Program stewardship.
15) Monitors the following:
    i) ERFO field activities.
    ii) ERFO funds request and Program of Projects.
    iii) Obligation of ERFO funds.
16) Notifies each applicant:
    i) When the $100,000,000 cap per State per disaster has been exceeded or met.
    ii) How ERFO funds will be transferred.
    iii) When Congress has waived the $100,000,000 cap per State per disaster.
17) Provides:
    i) Status and Fiscal Year Accomplishment reports, as required.

ii) Technical assistance and training.
iii) Status of the ERFO Program to the ER Program Manager.
iv) Guidance regarding administration of the ERFO Program.

18) Reinstates unobligated ERFO funds to applicants at the end of the fiscal year.
19) Reconciles fund transfers with the final Disaster Repair Closeout Report and letter.

d) Federal Lands Highway Director, Office of Program Development
1) Advises the applicant's headquarters office about the availability of funds.
2) Approves ERFO Policy changes and ERFO Program agreements.
3) Replies to congressional and public correspondence.
4) Decides appeals.

e) Associate Administrator, Office of Federal Lands Highway
1) Delegates Administration of the ERFO Program.
2) Decides appeals.

5.3. Other Agency Specific Roles and Responsibilities

a) Bureau of Indian Affairs notifies the Federal Lands Highway Division Engineer of damages sustained on Indian Reservation Roads. In addition, they are also required to inform the following entities of ERFO activities when such activities may impact:
1) Indian tribal governments.
2) State transportation agencies.
3) Counties.

b) Bureau of Land Management notifies the Federal Lands Highway Division Engineer about damages to Land Management Roads and Trails under the jurisdiction of a State or county providing access to or through public lands administered by the Bureau of Land Management.

c) County, Local Governments and Other Agencies:
1) Advises Federal Lands Management Agencies when damages are sustained on Federal Roads under their jurisdiction.
2) Enters into a Reimbursement Agreement with the Federal Lands Highway Division Engineer.
3) Has the same roles and responsibilities as an applicant as described in Section 5.1 (a).

d) Forest Service notifies the Federal Lands Highway Division Engineer about damages to Forest Highways that are under Forest Service or county jurisdiction.

e) Indian Tribal Government
1) Notifies the serving Bureau of Indian Affairs agency or Regional Office that Indian Reservation Roads have been damaged and requests that the Bureau of Indian Affairs apply for ERFO funds to make repairs.
2) Participates in the following activities:
i) Damage Survey Team.
ii) Detailed site inspections.
iii) Final inspection of ERFO projects on/near their reservation(s).

f) State Transportation Department

1) Applies for ERFO funds through the Federal Lands Highway Division Engineer for all Forest Highways that are Federal-Aid Highways.
2) Advises the Federal Lands Highway Division Engineer and Federal Lands Management Agencies of damage sustained on Federal Land Management Agency roads.
3) Enters into Reimbursement Agreements with the Federal Lands Highway Division Engineer, prior to performing eligible ERFO repairs.
4) Has the same roles and responsibilities as an applicant as described in Section 5.1 (a).

5.4. Federal Highway Administration

The Secretary of the Department of Transportation has delegated the administration of the emergency relief provision contained in Title 23 United States Code, Section 125, to the Federal Highway Administrator.

APPENDIX A – DEFINITIONS

Abbreviated plans. - A set of construction drawings for a project, may not contain detailed plan and profile sheets, cross-section sheets, and culvert profile sheets. The drawings must provide all essential information necessary to describe the repairs to be accomplished and to allow bidders to properly estimate unit prices for the contract.

Accelerated Release. - A transfer of ERFO funds for emergency repairs made prior to the approval of the Program of Projects and after the positive finding is issued. These funds are only to be used by the applicant to make emergency repairs to re-establish its vital transportation system(s) after a catastrophic failure or natural disaster over a wide area.

Acknowledgement letter.- A letter issued by the Federal Lands Highway (FLH) Division Engineer (DE) to the applicant upon receipt of the applicant's Notice of Intent letter providing basic information about the subsequent disaster assessment, eligibility information, and application procedures, including the necessary coordination and cooperation between the applicant and FLH DE.

Administrative road. - A road under the jurisdiction of a federal agency or Native American Indian tribal government that does not meet the definition of a public road. These roads may be closed locally at any time for any reason and remain closed indefinitely. Some of these roads may be gated or closed by earth barricades.

Allocation. - An administrative distribution of funds for programs that do not have statutory distribution formulae. Allocations are made authorizing an amount of funds, by program code, for a specific ERFO disaster event.

Allotment. - Provides current year obligation authority to the FLH or federal agency (FHWA's Allotment Advice Form 370).

Applicant. - Any federal agency which submits an application for emergency relief (ERFO funds) and that has authority to repair or reconstruct federal roads. State transportation agencies, Native American Indian tribal governments, counties, and other local governments apply indirectly through a federal agency.

Betterment. - Added features/facilities, not existing prior to the natural disaster or catastrophic failure, to help prevent future damage.

BIA. - Bureau of Indian Affairs.

Bureau of Land Management Highways. - Designated public roads that provide access to or are within lands administered by the Bureau of Land Management.

Catastrophic failure.- The sudden failure of a major element or segment of a Federal road which is not primarily attributable to gradual and progressive deterioration or lack of proper maintenance. The closure of a facility because of imminent danger of collapse is not in itself a sudden failure.

CFLHD - Central Federal Lands Highway Division.

Comparable facility. - For roads, a facility identical in capacity and character to the original. For bridges, a facility which meets the current geometric and construction standards required for the type and volume of traffic that such a facility will carry over its design life.

Construction engineering. - The work performed to supervise the construction operations through inspection of construction repairs, payments to the contractor, and otherwise representing the interests of the government.

Construction limits. - The limits on each side of the project which establish the area disturbed by construction operations and beyond which no disturbance is permitted.

Corps recreation road. - A public road that provides access to or within U.S. Army Corps of Engineers recreation or administrative areas for which title and maintenance responsibility is vested in the U.S. government.

Damage Survey Report (DSR). - A FLH ERFO form that documents damage by road/route number and mile post, scope of work, estimated repair costs, and eligibility recommendations.

Damage site survey team. - team composed of Federal Highway Administration (FHWA) and applicant representatives responsible for conducting inspections of damage sites. They collect the information used to prepare the Damage Survey Report.

Decommissioning or Storing. - Decommissioning, including storage of the road, may include one or more of the following:

a) Blocking the entrance to a road and installing water bars;
b) Removing culverts, re-establishing drainage-ways; and
c) Removing unstable fills, pulling back road shoulders, and erosion control.

Defense access road. - Public road designated by the Military Transportation Management Command that provides primary access to a military reservation, defense industry, or Department of the Air Force missile silos.

Department of Energy roads. - Public roads that provide access to or within facilities such as transmission lines, switch yards, substations, and related facilities on public lands, including those lands and facilities located on withdrawn public lands, previously under the jurisdiction of the Bureau of Reclamation (i.e., owned by the Western Area Power Association).

Disaster assessment. - An on-site evaluation of a site or sites conducted by a Disaster Assessment Team to determine extent, type, impact of damage, and cost to repair damage.

Disaster assessment report. - Documentation of the extent, type, impact, and cost of damage to transportation facilities that is used to determine whether an event, a natural disaster or catastrophic failure, has occurred and includes recommendations for a Positive or Negative Finding to the FLH DE.

Disaster assessment team. - A team of FHWA and applicant representatives who conduct a Disaster Assessment.

Disaster number. - Reference number issued by the FLH ERFO Program Manager to the applicant indicating state, fiscal year of the disaster event (four digits only), sequential number of approved events for that state; and federal agency. For example, WA2010-2-NPS indicates that it is for the second disaster event in Washington during fiscal year 2010 for the National Park Service. It could be either the first or the second Emergency Relief (ER) or ERFO disaster event in Washington.

DSR. - See Damage Survey Report.

DSR number. - Number assigned to each Damage Survey Report by the applicant. The number may contain the road/route number and the mile post number.

EFLHD. - Eastern Federal Lands Highway Division.

Emergency repairs. - Those repairs, including necessary preliminary engineering, construction engineering, and temporary traffic operations, undertaken with 6 months after a natural disaster or catastrophic failure to:

a) Restore essential travel;
b) Protect remaining facilities, and
c) Minimize the extent of damage.

Engineered Surfaced. - Consisting of manufactured materials (asphalt, concrete or crushed aggregate material) that are firm and stable under all weather conditions.

Equipment. - Apparatus that is tangible, non-expendable, personal property having a useful life of more than one year and an acquisition cost of $5,000 or more per unit.

ERFO. - Emergency Relief for Federally Owned roads.

ERFO Coordinator. - An individual designated by the FLH DE to have overall ERFO Program coordination responsibility within the Division.

ERFO funds. - Emergency relief funds available for expenditure under the authority of 23 U.S.C. 125(a) and (c).

ERFO Program Manager. - An individual designated by the FLH Associate Administrator and the Director, Office of Program Development, who is the FLH point of contact for the administration and oversight of the nationwide ERFO Program.

Event. - See Natural Disaster and Catastrophic Failure.

Event number. - See Disaster Number.

Facility. Road or bridge.

Federal-aid billing system. - The financial accounting system used by the FHWA's Federal-aid Divisions to reimburse states for eligible federal-aid expenses.

Federal-aid division office. - FHWA field office located in every state.

Federal Lands Highways. - Forest Highways, Park Roads and Parkways, Refuge Roads, Indian Reservation Roads, and Public Lands Highways that are public roads.

Federal Lands Highway Division Engineer. - The individual designated to have overall responsibility for the operations of one of the three FLH Divisions.

Federal Lands Highway Associate Administrator. - The individual designated to have overall responsibility for the administration and delivery of the FLH Program.

Federal Lands Highway Director, Office of Program Development. The individual designated to have overall responsibility for the program development and oversight of the FLH Program.

Federal road. Forest Highways, Federal Lands Highways, Forest Development Roads, Public Lands Development Roads, Public Lands Highways, Park Roads and Parkways, Refuge Roads, and Indian Reservation Roads as defined under 23 U.S.C. 101(a).

FHWA. - Federal Highway Administration.

Finding. - A letter or other official correspondence issued by the FLH DE to an applicant giving notification that, pursuant to 23 U.S.C. 125, federal roads have (positive finding) or have not (negative finding) been found to have suffered serious damage as the result of a natural disaster over a wide area, or a catastrophic failure.

FLH. - Federal Lands Highway.

FLHP. - Federal Lands Highway Program.

Force account. - A method of reimbursement for repair work performed by the applicant that is based on actual labor, equipment, and material costs.

Forest development roads and trails. - Forest roads and trails under the jurisdiction of the Forest Service.

Forest highway. - A Forest Road under the jurisdiction of, and maintained by, a public authority and open to public travel.

Forest road. - A road wholly or partly within or adjacent to, and serving, the U.S. Forest Service that is necessary for the protection, administration, and utilization of the National Forest System and the use and development of its resources.

Functionally obsolete. - A bridge is functionally obsolete when the deck geometry, load carrying capacity (comparison of the original design load to the state legal load), clearance, or approach roadway alignment no longer meets the usual criteria for the system of which it is an integral part.

FY..- Fiscal year.

Heavy maintenance. - Work usually done to repair damage normally expected from seasonal and occasionally unusual natural conditions or occurrences is considered heavy maintenance. This includes work at a site required as a direct result of a disaster which can reasonably be accommodated by a state or local road authority's maintenance, emergency or contingency program even though it exceeds the minimum site threshold of $5,000. Repair work which normally includes the following:

a) Grading roads rutted by traffic;
b) Cleaning or reshaping ditches;
c) Correction of minor embankment or cut slope slip-outs;
d) Removal of minor deposits of debris on roadways;
e) Correction of minor erosion damage to traveled way, shoulders (i.e., loss of surfacing down to sub-grade), or embankments;
f) Removal of minor mud deposits on roadways;
g) Minor pavement settlement unless significantly accelerated by the disaster event;
h) Minor damage to guardrail;
i) Minor riprap damage;
j) Replacement of short, isolated sections of surfacing washed away by flood water;
k) Cleaning of culverts;
l) Minor damage to culverts;
m) Minor repair of fencing;
n) Repair or replacement of isolated traffic signs;
o) Area-wide projects to correct extensive minor damage; and
p) Specialized equipment not normally required by routine maintenance.

Indian reservation roads. - Public roads that are located within or provide access to an Indian Reservation or Indian trust land or restricted Indian land which is not subject to fee title alienation without the approval of the federal government, or American Indians and Alaska Native villages, groups, or communities in which American Indians and Alaskan Natives reside, whom the Secretary of the Interior has determined are eligible for services generally available to American Indians under federal laws specifically applicable to them.

Interagency agreement. - A Memorandum of Agreement (MOA) between the FHWA and a federal agency establishing procedures for ensuring full cooperation under the ERFO

Program and the administration of ERFO funds to repair or reconstruct agencies' federally-owned roads damaged by natural disasters or catastrophic events.

Life cycle cost analysis. - A systematic quantitative method of assessing the desirability of adding Betterment to an ERFO project. The analysis must be based solely on discounted future ERFO eligible repair costs (both with and without the betterment), and the present cost of the betterment. The analysis cannot include other factors typically included in highway life cycle cost evaluations, such as traffic delay costs, added user costs, motorist safety, maintenance, and economic impacts.

Major loss. - A loss of materials from a roadway (e.g. gravel surfacing) that results in the operation or usefulness of the roadway being severely diminished to the extent that the roadway is not traversable by all classes of vehicles or the roadway is no longer safe to travel.

Military installation roads. - A public road that provides access within military installations for which title and maintenance responsibility is vested in the U.S. government.

Multi-use trails. - Provides individuals with mobility impairments to safely access and share a portion of an all weather engineered trail surface (asphalt, concrete or crushed aggregate material) with other recreational users.

National Environmental Policy Act (NEPA).- A law enacted in 1969 that set the national environmental policy, established the basis for environmental impact statements, and created the Council on Environmental Quality. NEPA also requires federal agencies to use an interdisciplinary approach in planning and decision making for actions that impact the environment, and the preparation of an environmental impact statement on all major federal actions significantly affecting the human environment.

Natural disaster.- A sudden and unusual natural occurrence such as a flood, hurricane, severe storm, tidal wave, earthquake, or landslide that causes serious damage over a wide geographical area.

Notice of Intent letter. – First formal application for ERFO funds submitted by the applicant to the FLH DE (see Appendix C.1).

Open to public travel. - A road section that is passable by standard passenger vehicles and open to the general public for use without restrictive gates, prohibitive signs, or regulation other than restrictions based on size, weight, or class of registration.

Parking lots. - Areas that are all weathered engineered surfaced (asphalt, concrete or crushed aggregate material) at locations designated for parking vehicles and staging of recreational equipment providing access to significant points of interest.

POP.- Program of Projects.

Park road. - A public road, including a bridge built primarily for pedestrian use but with capacity for use by emergency vehicles, that is located within, or provides access to, an area in the National Park System with title and maintenance responsibilities vested in the U.S. government.

Parkway. - A public road authorized by an Act of Congress on lands to which title is vested in the U.S. government.

Permanent work (permanent repair). - Repair or reconstruction to pre-disaster or other allowed geometric and construction standards and related PE and CE in accordance with an approved DSR.

Preliminary engineering (PE). - The preparation of plans, specifications, and estimates and other related documentation for use to construct repairs.

Program code. An alpha-numeric code used by the FHWA and the applicant to track the obligation and expenditure of ERFO funds.

Program of Projects (POP). - A list of projects either proposed by the applicant or approved by the FLH DE. The approved Program of Projects establishes the maximum amount of ERFO funds to be distributed pending availability.

Programmed project. - A project in which the construction phase is included in the FLH Program or State Transportation Improvement Plan, or in which the contract plans are being prepared for advertisement within the next three years.

Public authority. - A federal, state, county, town or township, American Indian tribe, municipal or other local government or instrumentality with authority to finance, build, operate, or maintain toll or toll-free facilities.

Public lands development roads. - Roads under the control of the Secretary of the Interior that the Secretary determines are of primary importance for the development, protection, administration, and utilization of public lands and resources.

Public lands highway. - A forest road under the jurisdiction of, and maintained by, a public authority and open to public travel; or any highway through unappropriated or unreserved public lands, non-taxable Indian lands, or other federal reservations under the jurisdiction of, and maintained by, a public authority and open to public travel.

Public road. - Any road or street under the jurisdiction of and maintained by a public authority and open to public travel.

Refuge road. - Public roads owned by the Service on national wildlife refuges, national fish hatcheries, and other administrative units such as the National Conservation Training Center, meet the definition of Public Lands Highways.

Replacement facilities. - Completely reconstructed (vs. repaired) facility built to the following standards:

a) Bridges built to meet current geometric and construction standards required for the type and volume of traffic is will carry during its design life.
b) Roads built to current design standards of comparable capacity and character to the destroyed facility.

Right-of-way. - Land acquired for and devoted to transportation purposes.

Road Management Objectives (RMOs).- The U.S. Forest Service's official forest plan implementation documents that display the intent and purpose of each road to provide access for the implementation of land and resource management plans. RMOs must be attested to as accurate by a line officer, usually the U.S. Forest Service District Ranger.

Roadway cross-section. - The finished or proposed finished road section between construction limits.

Serious damage. - Heavy, major, or unusual damage to a highway which severely impairs the safety, or usefulness of the highway or results in road closure and must be beyond the scope of heavy maintenance.

Slides. The removal of rock and mud slides from a Federal road either associated with the overall natural disaster or by itself to provide a safe roadway.

State transportation agency. - The department, commission, board, or official of any state charged by its laws with the responsibility for transportation facilities.

Statewide Transportation Improvement Program (STIP). - A staged, multi-year, statewide, intermodal program (list) of transportation projects consistent with the

Statewide Transportation Plan including the long-term vision of the state's multi-modal and intermodal transportation system.

Structurally deficient. - A bridge is structurally deficient when there is an extreme restriction on its load-carrying capacity, or it is frequently over-topped during flooding events causing severe traffic delays.

Trail. - A general term for a route used exclusively for travel by foot, or trail vehicle (motorized or non-motorized).

Transportation Improvement Program (TIP). - A staged, multi-year, metropolitan, intermodal program (list) of transportation projects that is consistent with the short-term and long-term vision of the metropolitan area's intermodal transportation system.

Traveled way. The portion of the roadway for the movement of vehicles, exclusive of the shoulders.

Western Area Power Association. - Department of Energy public roads that provide access to western-owned facilities such as transmission lines, switch yards, substations, and related facilities on public lands, including those lands and facilities located on withdrawn public lands, previously under the jurisdiction of the Bureau of Reclamation.

WFLHD. - Western Federal Lands Highway Division.

Appendix B – ERFO Disaster Assistance Service Area Reference Matrix

ERFO Disaster Assistance Service Area Reference Matrix						
State	Federal Lands Management Agency/Servicing Federal Lands Highway Division Office					
	Bureau of Indian Affairs Department of Defense Army Corps of Engineers	Bureau of Land Management	Forest Service	Bureau of Reclamation	Fish and Wildlife Service	National Park Service
Alabama	EFLHD		EFLHD		EFLHD	EFLHD
Alaska	WFLHD	WFLHD	WFLHD	WFLHD	WFLHD	WFLHD
Arizona	CFLHD	CFLHD	CFLHD	CFLHD	CFLHD	CFLHD
Arkansas	EFLHD		EFLHD		EFLHD	EFLHD
California	CFLHD	CFLHD	CFLHD	CFLHD	CFLHD	CFLHD
Colorado	CFLHD	CFLHD	CFLHD	CFLHD	CFLHD	CFLHD
Connecticut	EFLHD		EFLHD		EFLHD	EFLHD
Delaware	EFLHD		EFLHD		EFLHD	EFLHD
Florida	EFLHD		EFLHD		EFLHD	EFLHD
Georgia	EFLHD		EFLHD		EFLHD	EFLHD
Hawaii	CFLHD	CFLHD	CFLHD	CFLHD	CFLHD	CFLHD
Idaho	WFLHD	WFLHD	WFLHD	WFLHD	WFLHD	WFLHD
Illinois	EFLHD		EFLHD		EFLHD	EFLHD
Indiana	EFLHD		EFLHD		EFLHD	EFLHD
Iowa	EFLHD		EFLHD		EFLHD	EFLHD
Kansas	CFLHD		EFLHD	CFLHD	EFLHD	CFLHD
Kentucky	EFLHD		EFLHD		EFLHD	EFLHD
Louisiana	EFLHD		EFLHD		EFLHD	EFLHD
Maine	EFLHD		EFLHD		EFLHD	EFLHD
Maryland	EFLHD		EFLHD		EFLHD	EFLHD
Massachusetts	EFLHD		EFLHD		EFLHD	EFLHD

ERFO Disaster Assistance Service Area Reference Matrix						
State	Federal Lands Management Agency/Servicing Federal Lands Highway Division Office					
	Bureau of Indian Affairs Department of Defense Army Corps of Engineers	Bureau of Land Management	Forest Service	Bureau of Reclamation	Fish and Wildlife Service	National Park Service
Michigan	EFLHD		EFLHD		EFLHD	EFLHD
Minnesota	EFLHD		EFLHD		EFLHD	EFLHD
Mississippi	EFLHD		EFLHD		EFLHD	EFLHD
Missouri	EFLHD		EFLHD		EFLHD	EFLHD
Montana	WFLHD	WFLHD	WFLHD	WFLHD	WFLHD	WFLHD
Nebraska	CFLHD	CFLHD	CFLHD	CFLHD	CFLHD	CFLHD
Nevada	CFLHD	CFLHD	CFLHD	CFLHD	CFLHD	CFLHD
New Hampshire	EFLHD		EFLHD		EFLHD	EFLHD
New Jersey	EFLHD		EFLHD		EFLHD	EFLHD
New Mexico	CFLHD	CFLHD	CFLHD	CFLHD	CFLHD	CFLHD
New York	EFLHD		EFLHD		EFLHD	EFLHD
North Carolina	EFLHD		EFLHD		EFLHD	EFLHD
North Dakota	CFLHD	CFLHD	CFLHD	CFLHD	CFLHD	CFLHD
Ohio	EFLHD		EFLHD		EFLHD	EFLHD
Oklahoma	CFLHD		EFLHD	CFLHD	EFLHD	CFLHD
Oregon	WFLHD	WFLHD	WFLHD	WFLHD	WFLHD	WFLHD
Pennsylvania	EFLHD		EFLHD		EFLHD	EFLHD
Puerto Rico	EFLHD		EFLHD		EFLHD	EFLHD
Rhode Island	EFLHD		EFLHD		EFLHD	EFLHD
South Carolina	EFLHD		EFLHD		EFLHD	EFLHD
South Dakota	CFLHD	CFLHD	CFLHD	CFLHD	CFLHD	CFLHD
Tennessee	EFLHD		EFLHD		EFLHD	EFLHD
Texas	CFLHD		EFLHD	CFLHD	CFLHD	CFLHD
Utah	CFLHD	CFLHD	CFLHD	CFLHD	CFLHD	CFLHD
Vermont	EFLHD		EFLHD		EFLHD	EFLHD
Virginia	EFLHD		EFLHD		EFLHD	EFLHD
Virgin Islands	EFLHD		EFLHD		EFLHD	EFLHD
Washington	WFLHD	WFLHD	WFLHD	WFLHD	WFLHD	WFLHD
Washington, DC	EFLHD		EFLHD		EFLHD	EFLHD
West Virginia	EFLHD		EFLHD		EFLHD	EFLHD
Wisconsin	EFLHD		EFLHD		EFLHD	EFLHD
Wyoming Yellowstone Grand Tetons	CFLHD	CFLHD	CFLHD	CFLHD	CFLHD	CFLHD WFLHD WFLHD

Legend		
EFLHD Eastern Federal Lands Highway Division 21400 Ridgetop Circle Sterling, VA 20166 Phone: (703) 404- 6201	CFLHD Central Federal Lands Highway Division 12300 West Dakota Ave. Lakewood, CO 80228 Phone: (720) 963-3550	WFLHD Western Federal Lands Highway Division 610 East Fifth Street Vancouver, Washington 98661-3801 Phone: (360) 619-7700

APPENDIX C – CORRESPONDENCE (SAMPLES, INSTRUCTIONS, TABLES AND REPORTS)

C.1. Notice of Intent

Letter to be written by the applicant

Date

Division Engineer
Federal Highway Administration
[Division name]
[Address]
[City, State, Zip]

This letter is our formal Notice of Intent to apply for emergency relief funds to assist with the needed repairs and reconstruction under provisions of 23 U.S.C. 125. On [date], we conveyed by telephone to [name] our informal Notice of Intent to seek emergency relief funds.

We request a finding on the eligibility of damage to roads and trails that occurred on the [Forest, District, Park, Reservation] in the Counties of [name] in the state of [name].

The damage occurred on [specific date(s)].

Damage included failure over a wide area of: [*number*] bridges

[*number*] bridge approaches

[*number*] roads

The damage was caused by [*describe the nature and intensity of the disaster event or catastrophic failure*].

The preliminary cost estimate to restore eligible facilities to their pre-event condition is: [*list or state dollar amounts*]

Counties covered by a Presidential Disaster Declaration include [*list*]. Counties covered by a Governor's Proclamation of Emergency include [*list*].

Please ensure that responses to this letter are copied to [name, title, address]. If you have any questions, please contact [name, title] at [telephone number].

Sincerely,

Signature
____

[Name]
[Title]

C.1.1. Notice of Intent letter instructions

First formal application for ERFO funds submitted by the applicant to the FLH DE. (*Special note: Please include the FLH ERFO coordinator on your distribution list.)* Components of the letter should include the following:

a) Statement of intention to apply for Emergency Relief
b) Request for a finding
c) Location of damage
d) Date(s) of occurrence
e) Types of damage
f) Estimate of repair/reconstruction costs
g) General indication of the extra-ordinary character of the natural disturbance
h) Indication of the extent of areas affected outside the land(s) administered by the applicant
i) Counties covered by U.S. Presidential Disaster Declaration made under the Robert T. Stafford Disaster Relief and Emergency Assistance Act (42 U.S.C. 5121 et seq.)

C.2. Acknowledgment Letter

To be written by FLH DE

Date

[Applicant] [Address]
[City, State, Zip]
Re:/Subject: Emergency Relief for Federally Owned Roads
Acknowledgment Letter
Event Name (2010 June Flooding)
Region/Unit, State

Dear Applicant:

This letter is to acknowledge receipt of your Notice of Intent dated [date] to apply for Emergency Relief for Federally Owned roads (ERFO) funds under 23 U.S.C. 125 for the repair of roads and trails on the [Forest, District, Park, Reservation].

You should proceed with the performance of the emergency operations necessary to restore essential travel; to protect the remaining facilities; to reduce the extent of damage; to begin preliminary engineering (consisting of surveys, designs, and permanent restoration work that is required as an associated part of the emergency operations) and to use agency forces and/or negotiated equipment rental contracts as necessary to perform the work.

To maintain eligibility of repairs for ERFO funds, you must (1) photograph each site prior to making any repairs to document the extent of damage, and (2) track emergency repair expenditures by damage site.

The eligibility of such repairs for ERFO funds will be contingent upon the issuance of a *positive finding* on the eligibility of the disaster and subsequent approval of the repairs by Western [Central, Eastern] Federal Lands Highway Division, Federal Highway Administration.

The basis for the decision will be a Disaster Assessment Report with estimates of cost to repair or reconstruct the damaged routes and (evidence of wide area damage and intensity of the disaster event) or (evidence the failure occurred on a major segment of the road system causing a disastrous impact). [Since your estimated damage is less than the current threshold, other ERFO eligible Federal Land Management Agency or Tribe damage from the same disaster event must meet or exceed the current threshold of $700,000.]

Submit all information listed in Paragraph 2.1.2a of the ERFO Disaster Assistance Manual.

Upon receipt, a representative from this office will contact your field unit representative to make arrangements to review the disaster event damage to determine eligibility and begin preparing the Disaster Assessment Report. If a *positive finding* is made, we will make further arrangements for individual site inspections (if required) to prepare Damage Survey Reports.

You have the option to request the Accelerated Release of ERFO funds which may only be used to offset (reimburse) the eligible costs to agencies for making emergency repairs to re-establish its vital transportation system(s). These funds, subject to availability, may be only transferred after a *positive finding* has been issued.

If you have questions or need further information, please contact the Federal Lands Highway Division ERFO coordinator, [name] at [telephone number].

Sincerely,

Signature

[*Name*]
[*Title*]

Cc: Local FHWA div. office; FLH ERFO Program Manager; FHWA ER Program Manager; superintendent of field unit (cc list from notice of intent letter) with copy of incoming to all.

C.3. Disaster Assessment Report

To be written by the Federal Lands Highway Division ERFO coordinator The Disaster Assessment Report should include the following elements:

a) The intensity of the natural disaster or catastrophic failure. In the case of flooding, the Disaster Assessment Report shall include available information regarding high water elevations, stream-gauge data, and rain-gauge data.
b) Description of the nature, cause, and seriousness of the failure (if a catastrophic failure). Explain if the catastrophic failure was caused by gradual and progressive deterioration or the lack of maintenance. If appropriate, give the names of parties that may have contributed to the failure. Also state if insurance, cost share, or other compensation will pay for any of the repairs.
c) Dates of the occurrence.

d) Extent of the affected area.
e) Description of the damage to federal roads and other roads (Federal-aid highways, state, county, and city).
f) Copies of Emergency Relief Reports and newspaper articles, as available.
g) Copies of Presidential Disaster Declarations and Governor Proclamations, as available.
h) Approximate cost of repairs or reconstruction of the damage.
i) Photographs of typical damage and appropriate maps showing the location of each area and extent of damage.
j) A discussion on the impact of the subject damage. For example, a destroyed bridge with limited traffic has little impact, but a bridge serving as the principal access to homes, large tracts of land, or carrying large amounts of traffic will have a significant impact.
k) A concluding statement indicating a natural disaster or catastrophic failure causing severe damage has (or has not) occurred to federal roads and makes a specific recommendation for a *finding*.

C.4. Positive Finding Letter

To be written by FLH DE

Date
[Applicant name] [Address]
[City, State, Zip]

Re:/Subject: Emergency Relief for Federally Owned Roads Positive Finding Letter (ERFO Disaster Number)

Dear Applicant:

Your letter dated [insert date], requested Emergency Relief for Federally Owned roads (ERFO) funds be made available for the repair of [FMLA Unit, State]. The dates of said damage are from [beginning date] to [end date].

An assessment of the [catastrophic failure/natural disaster] has been completed and pursuant to Title 23 U.S.C. Section 125, I find that Emergency Relief funds, subject to availability, will be provided for eligible repairs on the following [Forest, District, Park, Reservation].

The disaster identification number will be [ERFO disaster event number]. Please use this number as the reference for any correspondence.

Arrangements will be made by my staff to assist you in completing joint damage site inspections and reports.

Upon completion of the site inspections, your office must submit a Program of Projects within 90 days from receipt of this letter.

The Program of Projects must include detailed descriptions and estimated costs of emergency repairs and proposed permanent repairs at each damage site. Following its approval by this office, ERFO funds may be requested by your headquarters office in accordance with the established procedures.

ERFO work shall be given prompt attention and priority over non-emergency work and programmed projects.

Projects not under construction by the end of the second fiscal year following the year that the disaster occurred will be withdrawn from the approved Program of Projects unless suitable justification is provided by the applicant to warrant retention.

If you have questions or need further information, please contact the Federal Lands Highway ERFO Coordinator, [name] at [telephone number].

Sincerely yours,

Signature

[Name]
[Title]

Cc: Local FHWA div. office; FLH ERFO Program Manager; FHWA ER Program Manager; superintendent of field unit with copy of incoming it all.

C.5. Negative Finding Letter

To be written by the FLH DE

Date

[Applicant name] [Address]
[City, State, Zip]
Re:/Subject: Emergency Relief for Federally Owned Roads Negative Finding Letter
Event Name (2010 June Flooding)

Dear Applicant:

Your letter of [date] requested Emergency Relief for Federally Owned Roads (ERFO) funds are made available for the repair of roads on [FMLA Unit, State]. The dates of the damage are from [beginning date] to [ending date].

An assessment of damage has been completed. It was found that certain damage did occur; however, these damages are [give brief explanation such as: considered heavy maintenance; roads are ineligible; threshold not met; does not meet the criteria for serious damage].

If you disagree with this finding, you may appeal as outlined in the ERFO Manual.

If you have any questions or need further information, please contact the Federal Lands Highway ERFO Coordinator, [name] at [telephone number].

Sincerely,

Signature

[*Name*]
[*Title*]

Cc: Local FHWA Div. Office; Federal Lands Highway ERFO Program Manager;
Federal Highway Administration ER Program Manager; Superintendent of Field Unit with copy of incoming to all.

C.6. Request for Accelerated Release Letter

To be written by the applicant

Date

Division Engineer
Federal Highway Administration
[Division name]
[Address]
[City, State, Zip]

Re:/Subject: Emergency Relief for Federally Owned Roads Request for Accelerated Release Letter ERFO Disaster Number

We have received your letter of Positive Finding dated [date] and formally request Accelerated Release of ERFO funds to re-establish vital transportation systems as indicated in the table below.

Request for Accelerated Release Letter Table			
Applicant			
Emergency Repair	Location	Funds Requested	Justification
Example: Culvert Replacement	Mile Post 10, Rte 101	$150,000	Access to private residences culvert washed out
Example: Bridge Replacement	Mile Post 20, Rte 101	$150,000	Vital link serving local hospital.

If you have questions, please contact [name], [title], at [telephone number].

Sincerely,

Signature

[Name]
[Title]

(Special Note: cc: Federal Lands Highway ERFO Coordinator)

C.7. Accelerated Release Response Letter

To be written by FLH DE

Date
[Applicant name]
[Address]
[City, State, Zip]
Re:/Subject: Emergency Relief for Federally Owned Roads Accelerated Release Response Letter ERFO Disaster Number
Event Name (2010 June Flooding)

Dear Applicant:

Your Request for Accelerated Release has been (approved/disapproved) as indicated below. These funds (subject to availability) may only be used to offset (reimburse) the eligible costs to Agencies for making emergency repairs to re-establish vital transportation system(s).

Request for Accelerated Release Letter Table					
Applicant				Division Engineer	
Emergency Repair	Location	Funds Requested	Justification	Approved/ Disapproved	Reason
Example: Bridge Replacement	Mile Post 10, Rte 101	$150,000	Access to private residences culvert washed out	Disapproved	Alternate access available.
Example: Bridge Replacement	Mile Post 20, Rte 101	$150,000	Vital link serving local hospital.	Approved	

This letter provides advice on Accelerated Release of ERFO funds approval only and does not transfer obligation authority or cash to accomplish the approved repairs. Please advise your headquarters that funds, subject to availability, in the amount of [amount] may be formally requested from the Federal Lands Highway ERFO Program Manager.

If you have any questions, please contact [*name*] at [*telephone number*].

Sincerely,

Signature

[Name]
[Title]

Your Request for Accelerated Release has been (approved/disapproved) as indicated below. These funds (subject to availability) may only be used to offset (reimburse) the eligible costs to Agencies for making emergency repairs to re-establish vital transportation system(s).

cc: Local FHWA Div. Office; Federal Lands Highway ERFO Program Manager; Federal Highway Administration
ER Program Manager; Superintendent of Field Unit with copy of incoming to all.

C.8. Site Inspection Equipment List

Description	Applicant	Provider Federal Lands Highway
Two maps identifying designated applicant's Federal Roads which clearly depict the damage site.	X	
A copy of the current ERFO Manual.		X
Supply of Damage Survey Report forms.		X
Camera and method to identify the photographed location (e.g., marker and pad, small chalk board and chalk).	X	X
A measuring tape or electronic distance measuring device.	X	X
Clipboard, pencils, pens, stapler, and extra staples, calculator, paper, and straight edge.	X	X
Applicant-owned equipment rental rates (with operator rate included).	X	
Local market, private industry operator labor rates (with operator rate included).	X	
Tabulation of recent average unit contract prices for selected repairs.	X	
Local market, private industry equipment rental rates.	X	
Notebook computer as needed.	X	X
Global Positioning System (GPS) Equipment (optional)		X
Hard hats.	X	
Safety vests.	X	X
First-aid kit.	X	X
Traffic control devices (i.e., cones).	X	

(Continued)

Description	Applicant	Provider Federal Lands Highway
Steeled-toe boots (optional).	X	X
Communication equipment (telephone/radio).	X	X
Clinometer	X	X
Compass	X	X
Spot Light	X	X

C.9. Damage Survey Report

Add Operational and Objective Maintenance Level

C.9.1. Instructions

a) **Sheet no. ___ of ___:** Sheet numbers are entered to sequence the DSRs.
b) **DSR No.:** Applicant assigned number given to each DSR containing at least the route number, name of road, and mile post.
c) **Disaster No.**: The disaster number is provided in the *Positive Finding* letter.
d) **Applicant:** Enter the name of the field unit (National Forest, National Park, etc.) and Agency name.
e) **Unit**: Forest Service District Office or U.S. Corp of Engineer project.
f) **County:** Enter the name of the county in which the damage site is located.
g) **State:** Enter the two-letter abbreviation of the state in which the damage site is located.
h) **Inspection date:** Enter the inspection date.
i) **Location of damage:** Enter the route number, name of the road, and milepost of the damage site in miles or kilometers according to your agency's metric policy.
j) **ADT:** Enter the average daily traffic for the road.
k) **Bridge data:** Enter type of bridge and the bridge number.
l) **Road data:** Enter the road classification, traveled way width, and surface type and thickness; shoulder width, thickness, and surface type and thickness.
m) **Inventory route number:** Route number from FHWA or BIS road inventory.
n) **Latitude/longitude:** Geographical position of the site.
o) **Maintenance Level:** Operational maintenance level (Forest Service only).

FEDERAL HIGHWAY ADMINISTRATION, Federal Lands Highway DAMAGE SURVEY REPORT (Title 23, Federal-Aid System/Federal Domain)			Sheet No: 1 of DSR No: Disaster No:
Applicant: Agency:	County:	State:	Date:
Unit:	Region:		Maintenance Level:
Route Name: FHWA Inventory Route Number:	Milepost:	Latitude: Longitude:	ADT:

Road Info	Traveled Way Width:	Surface Type:	Thickness:	Bridge Type:
Shoulder Info	Shoulder Width:	Surface Type:	Thickness:	NBIS #:

Description and Cause of Damage:	Scope/Description of Repair:

COST ESTIMATE for EMERGENCY REPAIRS*

Quantity	Unit	Item Description	Unit Price	Cost

Proposed: ◯ Force Account ◯ Contract ◉ Neither Total Emergency Repairs

COST ESTIMATE for PERMANENT REPAIRS*

Quantity	Unit	Item Description	Unit Price	Cost

Subtotal Permanent Repairs (continuation sheet)

Proposed: ◯ Force Account ◯ Contract ◉ Neither Total Permanent Repairs

Identify Betterment, if any, and provide justification*	
	15% Preliminary Engineering $ 0 10% Construction Engineering $ 0 Other: Other: TOTAL ESTIMATED COST $ 0 (Emergency and Permanent Repairs)
Submitted By: (Name and Title)	Signature: Date:
Reviewed by : — Eligible / Ineligible: No Entry	Signature: Date:
Recommended by: — No Entry	Signature:

*Attach Supplemental Sheets if necessary

Submit by Email

d) Description and cause of damage:

1) Provide a clear description about how the damage occurred.
2) Describe the amount of material lost, noting the length, width, and depth.
3) Include the size and condition of the structures.

4) Note any signs of a preexisting condition for slides and slumps such as sealed pavement cracks, vegetation growing in slope cracks, horizontal drains. Note pre-disaster condition of culverts and roadside ditches.
5) Provide a sketch or cross-section drawing with dimensions that illustrates the damage and current roadway width.

e) **Scope of work:**
1) Describe both emergency and permanent repairs.
2) Provide a brief outline of alternate repair option(s), if considered.

f) **Cost estimate for emergency repairs:** Show all emergency repairs as well as temporary traffic control devices, warning signs, DSR preparation, NEPA, and erosion control.
1) **Quantity:** Enter the number of units of equipment, labor, or materials. Show backup calculations on supplemental sheets.
2) **Unit:** Enter the unit of measures.
3) **Description:** Enter the name of the equipment, labor, and materials.
4) **Unit price:** Enter the unit price for the item.
5) **Cost:** Calculate the COST by multiplying the QUANTITY by the UNIT PRICE.
6) **Force account:** Mark this box if the emergency repairs were done by the applicant's employees.
7) **Contract:** Mark this box if the emergency repairs were done by a contractor.
8) **Total emergency repairs:** Add up the costs for each item and enter the total.

g) **Cost estimate for permanent repairs:** Use the type of cost items that typically appear in competitively bid construction contracts.

Special Notes:

- Do not use units of equipment and labor or more than 10 percent of the permanent repair costs.
- Do not use lump sum for more than 25 percent of repair costs.

1) **Quantity:** Enter the number of units of equipment, labor, or materials.
2) **Unit:** Enter the unit of measure.
3) **Description:** Enter the name of the equipment, labor, and materials.
4) **Unit price:** Enter the unit price for the item.
5) **Cost:** Calculate the COST by multiplying the QUANTITY by the UNIT PRICE.
6) **Force account:** Mark this box if the permanent repairs will be done by the applicant's employees.
7) **Contract:** Mark this box if the permanent repairs will be done by a contractor.
8) **Total permanent repairs:** Total the costs.

h) **Identify betterment and provide justification in the block provided:** Add separate sheets as needed for life cycle cost analysis calculations.

i) **Preliminary engineering:** Calculate the cost of preliminary engineering as a percentage of the total permanent repair cost. Average preliminary engineering as a percentage of the total permanent repair cost. Average preliminary engineering costs are 15 percent.

j) **Construction engineering:** Calculate the cost of construction engineering as a percentage of the total permanent repair cost. Average construction engineering costs are 10 percent.
k) **Right-of-way:** Enter the cost of acquiring the right-of-way that is necessary to make the repairs. Consider using construction easements whenever possible.
l) **Other:** Costs other than previously identified.
m) **Total estimated cost:** Enter the total cost for emergency repairs, permanent repairs, preliminary and construction engineering, and right-of-way. If there is cost sharing on the road, adjust the total amount to reflect the government's share. ERFO funds can pay only for the government's share of the repair costs.
n) **Submitted by:** Fill in the applicant's field unit, ERFO coordinator's submitter's name, title, signature, and date.

aa) **Reviewed by:** Reserved for use by Federal Lands Highway field reviewer. Enter reviewer's name and title; note your opinion of eligibility/ineligibility by marking the appropriate box; sign and date.
bb) **Recommended by:** Reserved for use by Federal Lands Highway Division ERFO coordinator. Enter the coordinator's name and title; note your opinion of eligibility/ineligibility by marking the appropriate box; sign and date.
cc) **Combining sites:** One or more eligible damage sites can be reported on one Damage Survey Report (DSR); however, large slides and bridges may not be grouped on one report. Use supplemental sheets as needed.

When multiple sites are combined on one DSR, provide separate supplemental sheets for each site. For each site, provide a description of damage, scope of repairs, and cost estimate. Summarize the cost by site on the first page of the DSR.

C.9.2. Supplemental Sheets

a) Duplicate the information contained on the original DSR about the site(s) at the top of each supplemental sheet.
b) The following items are required and should be attached to supplemental sheets:
    1) Additional explanations of the damage, scope of work, estimated costs of repairs, betterment justification, etc.
    2) Sketches or drawings illustrating the damage or proposed repairs.
    3) Photographs of the damage site(s).
    4) Map(s) showing the site location(s).
    5) Backup calculations for major construction item quantities.

C.10. Applicant's Program of Projects Letter

To be written by applicant

Date
Division Engineer
Federal Highway Administration
[Division name]
[Address]
[City, State, Zip]

Re:/Subject: Emergency Relief for Federally Owned Roads Program of Projects
Request Letter No: (number) ERFO Disaster Number
Event Name (2010 June Flooding)

Dear [Division Engineer name]*:*

We request *(initial/revised)* Program of Projects approval for roads damaged in disaster [disaster number] which occurred on [date] as a result of a *(natural disaster/catastrophic failure)* under our jurisdiction as summarized below.

Attached is our *(initial/revised)* proposed Program of Project's spreadsheet. *(Submit the original Damage Survey Reports if not previously submitted.)*

We request that the [Division name] Federal Lands Highway Division perform the following work *(includes description and site location.)*:

Program of Projects Summary Table				
Location of Damage	Organization Performing the Repairs	Previously Approved Amount	Requested Amount	Total Requested Amount
Mt Rainier	NPS	$0	$468,949	$468,949
	WFLHD	$0	$176,000	$176,000
N. Cascades				
Subtotal	NPS	$0	$468,949	$468,949
	WFLHD	$0	$176,000	$176,000
Grand Total		$0	$644,949	$644,949

If you have any further questions, please contact [name], at [telephone number].

Sincerely,

Signature

[Name]
[Title]

Enclosure: Program of Projects Table

(Special Note: Please include the Federal Lands Highway ERFO Coordinator on your distribution list.)

C.11. Program of Projects Table

Program of Projects Table 1 Date:							
Disaster Number: AZ2006-1-FS		Agency Sub_Unit: Coronado National Forest					
DAMAGE SURVEY REPORT		DESCRIPTION OF REPAIR	PROGRAM DOLLAR AMOUNT				PERMANENT REPAIR METHOD
Route Number	MP	Bold/Italic row(s) affected by change	Applicant	Federal Lands Highway Division	Other includes Tribe, State,	Total	Force Account/ Contract/ Other
				Office	Agency, County		
100A-1	0.00	Repair approach, curb, and ditchline	$14,581	$0	$0	$14,581	Contract
5-1	0.00	Repair LWC approaches	$11,261	$0	$0	$11,261	Contract
100-1	0	Replace fill, repair bridge in kind	$92,148	$0	$0	$92,148	Force Account
100-2	1	Repairs roadway, shoulders and ditch lines	$20,423	$0	$0	$20,423	Contract
100-10	0	MSE wall, regrade, repave	$307,619	$0	$0	$307,619	Contract
APPROVED PROGRAM TOTAL			$446,032	$0	$0	$446,032	
Revision/Change(s)							
100A-1	0.00	Repair approach, curb, and ditchline	$0	$0	$0	$0	Contract
5-1	0.00	Repair LWC approaches	$0	$0	$0	$0	Contract
100-1	0	Replace fill, repair bridge in kind	$0	$0	$0	$0	Force Account
100-2	1	Repairs roadway, shoulders and ditch lines	$0	$0	$0	$0	Contract
100-10	0	MSE wall, regrade, repave	$0	$0	$0	$0	Contract
APPROVED PROGRAM TOTAL			$0	$0	$0	$0	
Revised Program of Projects							
100A-1	0.00	Repair approach, curb, and ditchline	$14,581	$0	$0	$14,581	Contract
5-1	0.00	Repair LWC approaches	$11,261	$0	$0	$11,261	Contract
100-1	0	Replace fill, repair bridge in kind	$92,148	$0	$0	$92,148	Force Account
100-2	1	Repairs roadway, shoulders and ditch lines	$20,423	$0	$0	$20,423	Contract
100-10	0	MSE wall, regrade, repave	$307,619	$0	$0	$307,619	Contract
APPROVED PROGRAM TOTAL			$446,032	$0	$0	$446,032	

C.12. Program of Projects Approval Letter

To be written by the Federal Lands Highway

Date
[Applicant name] [Address]
[City, State, Zip]

Re:/Subject: Emergency Relief for Federally Owned Roads Program of Projects Approval Letter No: (number) ERFO Disaster Number
Event Name (2010 June Flooding)

Dear Applicant:

The proposed Program of Projects submitted with your letter dated [date], is approved as indicated below.

We are pleased to advise you that Federal Lands Highway is available to assist your agency per your request to perform work dated [date]. A representative from our office will be in contact with your field unit representative to coordinate this effort. Federal Highway Administration approval is required at stages *Approval of Plans, Specifications, and Estimates packages; Concurrence in rejection of bids; and Concurrence in award.*

Program of Projects Summary Table				
Location of Damage	Organization Performing the Repairs	Previously Approved Amount	Current Approved Amount	Total Approved Amount
Mt Rainier	NPS	$0	$468,949	$468,949
	WFLHD	$0	$176,000	$176,000
N. Cascades				
Subtotal	NPS	$0	$468,949	$468,949
	WFLHD	$0	$176,000	$176,000
Grand Total		$0	$644,949	$644,949

This letter provides advice on program approval only and does not transfer obligation authority or cash to accomplish approved repairs. Funds, subject to availability, in the amount of [amount] may be requested on an asneeded/written request basis. Please advise your office headquarters to formally request these funds from the Federal Lands Highway ERFO Program Manager. [*Please note: Transfer of ERFO funds will be withheld if prior year Disaster Status Reports have not been submitted to the Federal Lands Highway Division Engineer.]*

We will be requesting our Washington Headquarters to arrange the transfer of obligation authority on an as needed/written request basis in the amount of [amount] to [Division name] Federal Lands Highway Division.

ERFO work shall be given prompt attention and priority over non-emergency work and programmed projects. Projects not under construction by the end of fiscal year XXXX will be withdrawn from the approved POP unless a suitable justification is provided to warrant a time extension.

If you have questions or need further information, please contact the Federal Lands Highway Division ERFO Coordinator, [name], at [telephone number].

Sincerely,

Signature

[*Name*]
[*Title*]

cc:Local FHWA Div. Office; Federal Lands Highway ERFO Program Manager; Federal Highway Administration ER Program Manager; Superintendent of Field Unit with copy of incoming to all.

C.13. Request for ERFO Funds

To be written by the applicant

Date
[Applicant name] [Address]
[City, State, Zip]
Federal Lands Highway Associate Administrator
Federal Highway Administration 400 Seventh Street SW – HFL-1 Washington, DC 20590

Re:/Subject: Emergency Relief for Federally Owned Roads Request for
ERFO Funds Letter No: (number) ERFO Disaster Number
Event Name (2010 June Flooding)
Dear (ERFO Program Manager):

This is to request ERFO funds in the amount of [dollars] for disaster [number] which will be obligated in fiscal year [year].

Thank you for your assistance in this matter. If there are any questions, please contact [name] at [telephone number].

Sincerely,

Signature

[Name]
[Title]

Enclosure: Annual ERFO Funds Request Table (see C.18)
Special Note: include HFPD-5 in the *cc:* lines

C.14.Transfer of ERFO Funds

To be written by the Federal Lands Highway ERFO

Date

HFPD5-[month and FY]-[Program Code][FLMA Letter No. sequence]*[Tracking No.]

Refer to: HFPD-5

[Applicant name] [Applicant Title] [FLMA]
[Address]
[City, State, Zip]

Re:/Subject: Emergency Relief for Federally Owned Roads Allocation of ERFO Funds Letter No: (number)

Dear Applicant:

The purpose of this letter is to provide to [FLMA] funds made available under the Emergency Relief for Federally Owned Roads (ERFO) Program.

Based upon information provided by your staff, this office has determined that a total of [amount] in [Program Code] needs to be allocated. Therefore, we are allocating [amount] in [Program Code] Contract Authority to the disasters shown in Table A.

(OPTIONAL PARAGRAPH—BUDGET AUTHORITY)

A Federal Highway Administration Advice of Funds and an 1151 Non-expenditure Transfer for the amount of the allocation will be processed by our Office of Budget and Finance. Please bear in mind that obligations cannot be incurred until the 1151 Non-expenditure Transfer has been posted by Treasury and all activity against this funding must be reported separately from Federal-aid Program funding on the SF-133.

(OPTIONAL PARAGRAPH—CONTRACT AUTHORITY)

The Federal Highway Administration (FHWA) Office of Budget will issue a completed FHWA Form 370 confirming the allocation of funding shortly. Please continue to report ERFO obligations on the SF-133, separately from other Highway Trust Funds.

If there are any questions, please contact [ERFO Program Manager] at [telephone number].

Sincerely yours,

Signature

[*Name*]
[*Title*]

Enclosure must include Table for both Divisions and FLMA's

Cc:Local FHWA Div. Office; Federal Lands Highway ERFO Program Manager; Federal Highway Administration ER Program Manager; Superintendent of Field Unit with copy of incoming to all.

C.15. Allocation Memorandum to a Federal Lands Highway Division

To be written by the ERFO Program Manager

HFPD5-[month and FY]-[Program Code][FLMA Letter No. sequence]*[Tracking No.]

Subject: *ACTION:* Emergency Relief for Federally Owned Roads (ERFO) Funding [Fiscal Year], Allocation
Date:
From: Director, Office of Program Development In Reply Refer To: [HFPD5]

To: Federal Lands Highway Division Engineers

This memorandum allocates [amount] to Disasters [ERFO disaster number] [amount] and [ERFO disaster number] [amount] for approved repairs in ERFO funds, PC-[code], as shown on the attached Table A. The accounting flex field information is as follows: Fund, [identification], Budget Year, 0000, and BPAC, [identification].

By copy of this memorandum, we are requesting the Office of Budget and Finance to make the appropriate adjustments in the Financial Management Information System for [contract or budget] authority and obligation limitation, in accordance with Table A.

If you have any questions regarding this memorandum, please contact [name], ERFO Program Engineer, at [telephone number].

Attachment
Special Note: Include ERFO Coordinators in "cc" lines.

C.16. Request for Time Extension

To be written by the Applicant

Date
Division Engineer
Federal Highway Administration
[Division name]
[Address]
[City, State, Zip]

Re:/Subject: Emergency Relief for Federally Owned Roads Time Extension Request Letter No: (number) ERFO Disaster Number Event Name (2010 June Flooding)

Dear [Division Engineer name]:

The [Applicant name] requests a [time] extension for approved ERFO site [site location] which resulted from disaster number [disaster number].

Please see the attached table for justification statement and dates of major milestones. If you have any further questions, please contact [name], [title], at [telephone number].

Sincerely,

Signature
[Name]
[Title]

Attachment

C.16.1 Request for Time Extension Table

To be written by the applicant

Damage Survey Report Number	Justification statement	Project Status and Timeline
Example: BLU 2300-4.9	Example: We were delayed from accessing this site because of damage to county road, Route 6, that occurred during the OR 2007-2-BIA disaster. The damage at M.P. 1.0 has severely damaged the bridge abutment and it will not be repaired until later this fall. Route 2300 is a well maintained road that is the primary access to the landfill.	Example: (Provide completion/planned completion dates for the following deliverables) Applicant's POP Letter: [date] Topographic Survey and Mapping: [date] Environmental Resource Surveys: [date] Preliminary Design: [date] Draft NEPA Document: [date] Biological Assessment/Opinion: [date] Final NEPA (ROD, FONSI, CE): [date] Final PS&E: [date] Contract Award: [date] Contract Completion: [date]

C.16.2. Instructions for Time Extension Table

a) Submit justification for each DSR
b) Provide description of why the project was not completed on time. Provide description of primary purpose and need of the road.
c) Provide dates of major project milestones (actual/projected) as listed below to demonstrate the project was advanced on a priority basis and that a time extension is justified
    1) POP Letter date (Applicant's first submittal)
    2) Topographic survey and mapping date
    3) Environmental survey completion dates
    4) Preliminary design date
    5) Engineering recommendation dates (Geotechnical, Hydraulic, Structures, etc.)
    6) Draft NEPA document date
    7) Biological Assessment and Opinion dates
    8) Final NEPA document (ROD, FONSI, CE) date
    9) Final PS&E date

10) Advertisement date
11) Construction award date
12) Construction completion date

C.17. Approval of Request for Time Extension

To be written by the Federal Lands Highway Division Engineer

Date
[Applicant name]
[Address]
[City, State, Zip]
Re:/Subject: Emergency Relief for Federally Owned Roads
Approval of Request for Time Extension Letter Number :____ ERFO Disaster Number
Event Name (2010 June Flooding)

Dear Applicant:

We received your letter dated [date] requesting a time extension for Emergency Relief for Federally Owned Roads (ERFO) projects damaged in the [disaster number] disaster.

Based upon the information you presented, approval/disapproval is indicated on the attached table. The projects must be under contract as indicated on the attached table in order to ensure continued eligibility of ERFO funds.

Sites not eligible for time extensions should be removed from future ERFO fund requests. Any expenditure of ERFO funds on these sites after the approved completion date is prohibited. In addition, expenditure of any other program/agency funds after that date will not be reimbursed under the ERFO Program.

If you have questions or need further information, please contact the Federal Lands Highway Division ERFO coordinator, [name] at [telephone number].

Sincerely,

Signature
____

[Name]
[Title]

Attachment

cc: Local FHWA Div. Office; ERFO Program Manager, Federal Highway Administration ER Program Manager, superintendent of field unit with copy of incoming to all.

C.17.1. Time Extension Table

To be written by the Federal Lands Highway Division Engineer

Damage Survey Report Number	New Approved Contract award Date	Rationale for Decision
Example: BLU 2300-4.9	*Example: 9/30/2012*	*Example: The time extension is approved based on another eligible ERFO disaster delaying the repair of this site. We expect that the repair of this site will be accomplished as soon as access across the county bridge is restored.*

C.18. Annual ERFO Funds Request

To be written by the Applicant

ANNUAL ERFO FUNDS REQUEST						
Agency Administering ERFO Repairs	Disaster Number	Reimbursement (Payback) Funds		(3) Amount of Funds needed for Ongoing Projects	(4) Amount of Funds needed for this Fiscal Year	Totals
		(1) Agency Funds	(2) Federal Lands Highway			
		$	$	$	$	$
		$	$	$	$	$
		$	$	$	$	$
		$	$	$	$	$
		$	$	$	$	$
Total		$	$	$	$	$
Prepared by: Date: Phone:						

1) Amount of Agency funds used on repairs to be repaid by ERFO.
2) Amount of Federal Lands Highway Program funds used on repairs to be repaid by ERFO.
3) Amount of ERFO funds needed for Construction Engineering (CE)/Contract Modifications (CM) for active/ongoing construction projects.
4) The amount of funds needed for design, advertisement, award, and construction engineering for new projects this fiscal year.

C.19. Request for Closeout

To be written by the Applicant

Date

Division Engineer
Federal Highway Administration
[Division name]
[Address]
[City, State, Zip]

Re:/Subject: Emergency Relief for Federally Owned Roads Request for
Closeout Letter Number :___ ERFO Disaster Number
Event Name (2010 June Flooding)

Dear [Division Engineer name]:

The [Applicant name] requests a [final/partial] closeout of approved ERFO site(s) which resulted from disaster number [number]. The site(s), shown on the attached Disaster Closeout Report, have been repaired according to the description of repairs on the most recent DSR and the approved Program of Projects (except as noted). The total expenditure to repair all damage was [amount].

If you have any further questions, please contact [name], [title], at [telephone number].

Sincerely,

Signature ____________________

[Name]
[Title]

Attachment: Disaster Status Report.

C.20. Disaster Status Report

To be written by the Applicant

Note: As a status report, this report is due to the ERFO coordinator by October15 of each year until the disaster is closed out. If reports are not received, funding may be withheld.

As a closeout report, this report is due to the ERFO coordinator at the time of closeout submittal.

DISASTER STATUS REPORT

Date: ____________ Disaster Number: ____________ Sub-Unit: ____________

LOCATION DAMAGE SURVEY REPORT NUMBER	DESCRIPTION OF APPROVED REPAIR	APPROVED ESTIMATED COSTS				CURRENT OBLIGATIONS/ FINAL EXPENDITURES				(Anticipated/ Actual) Contract Award Date	(Anticipated/ Actual) Construction Completion Date	Date Closed
		Applicant	Federal Lands Highway Division	Other	TOTAL	Applicant	Federal Lands Highway Division	Other	TOTAL			
Example:BL U 2300-4.9	Reconstruct roadway embankment and replace two culverts	$56,000			$56,000					9/24/1997	10/30/2003	
TOTAL												
Example		$56,000			$56,000					9/24/1997	10/30/2003	

The submitted expenditures are the actual costs incurred to complete only approved eligible repairs. Attach a separate sheet to explain any deviations.

Signed: ………………………………………..Date: ………………………………….

C.21. Closeout Letter

To be written by the Federal Lands Highway Division Engineer

Date
[Applicant name]
[Address]
[City, State, Zip]

Re:/Subject: Emergency Relief for Federally Owned Roads
Program of Projects Letter [number] (*Partial/Final*) Disaster Closeout ERFO Disaster Number
Event Name (2010 June Flooding)

Dear Applicant:

We received your notification that (*all/a portion of the*) eligible ERFO activities associated with [disaster number] have been completed.

The revised total program and authorized expenses for eligible ERFO repairs for this disaster is [amount]. The previously approved program amount of [amount] was (*Overrun/Underrun*) by [amount].

Program of Projects Summary Table				
Location of Damage	Organization Performing the Repairs	Previously Approved Amount	Current Approved Amount	Total Approved Amount
Mt Rainier	NPS	$0	$468,949	$468,949
	WFLHD	$0	$176,000	$176,000
N. Cascades				
Subtotal	NPS	$0	$468,949	$468,949
	WFLHD	$0	$176,000	$176,000
Grand Total		$0	$644,949	$644,949

We have conducted a final review of the damage repair and supporting records and have discussed our findings with [name] from your office on [date].

As a result of our review, we find all repairs to be accomplished in reasonably close conformity to the approved program.

We hereby grant final acceptance of the subject damage repairs.

The expertise and diligence of your office and field personnel in restoring damaged roadways to stable and serviceable conditions through innovative and cost-savings techniques are appreciated.

By copy of this letter we are advising our Washington Headquarters that the subject ERFO program [disaster number] has received final inspection and acceptance at the final approved program amount of [amount] and should begin final fund reconciliation with your headquarters. If you have any questions, please contact [name] at [telephone number].

Sincerely,

Signature ________

[*Name*]
[*Title*]

cc: Local FHWA Div. Office; Federal Lands Highway ERFO Program Manager; Federal Highway Administration ER Program Manager; Superintendent of Field Unit with copy of incoming to all.

Appendix D – Law and Regulation

D.1. Select Provisions of 23 U.S.C., Highways, Related to the ERFO Program

Sec. 101. Definitions.

a) Definitions. In this title, the following definitions apply:
   1) County. The term "county" includes corresponding units of government under any other name in States that do not have county organizations and, in those States in which the county government does not have jurisdiction over highways, any local government unit vested with jurisdiction over local highways.
   2) Federal-aid highway. The term "Federal-aid highway" means a highway eligible for assistance under this chapter other than a highway classified as a local road or rural minor collector.
   3) Federal-aid system. The term "Federal-aid system" means any of the Federal-aid highway systems described in section 103.
   4) Federal lands highway. The term "Federal lands highway" means a forest highway, public lands highway, park road, parkway, refuge road, and Indian reservation road that is a public road.
   5) Forest development roads and trails. The term "forest development roads and trails" means forest roads and trails under the jurisdiction of the Forest Service.
   6) Forest highway. The term "forest highway" means a forest road under the jurisdiction of, and maintained by, a public authority and open to public travel.
   7) Forest road or trail. The term "forest road or trail" means a road or trail wholly or partly within, or adjacent to, and serving the National Forest System that is

necessary for the protection, administration, and utilization of the National Forest System and the use and development of its resources.

8) Highway. The term "highway" includes (A) a road, street, and parkway; (B) a right-of-way, bridge, railroad-highway crossing, tunnel, drainage structure, sign, guardrail, and protective structure, in connection with a highway; and (C) a portion of any interstate or international bridge or tunnel and the approaches thereto, the cost ofwhich is assumed by a State transportation department, including such facilities as may be required by the United States Customs and Immigration Services in connection with the operation of an international bridge or tunnel.
9) Indian reservation road. The term "Indian reservation road" means a public road that is located within or provides access to an Indian reservation or Indian trust land or restricted Indian land that is not subject to fee title alienation without the approval of the Federal Government, or Indian and Alaska Native villages, groups, or communities in which Indians and Alaskan Natives reside, whom the Secretary of the Interior has determined are eligible for services generally available to Indians under Federal laws specifically applicable to Indians.
10) Maintenance. The term "maintenance" means the preservation of the entire highway, including surface, shoulders, roadsides, structures, and such traffic-control devices as are necessary for safe and efficient utilization of the highway.

b) Park road. The term "park road" means a public road, including a bridge built primarily for pedestrian use, but with capacity for use by emergency vehicles, that is located within, or provides access to, an area in the National Park System with title and maintenance responsibilities vested in the United States.
c) Parkway. The term "parkway", as used in Section 2 of this title, means a parkway authorized by Act of Congress on lands to which title is vested in the United States.
d) Public authority. The term "public authority" means a Federal, State, county, town, or township, Indian tribe, municipal or other local government or instrumentality with authority to finance, build, operate, or maintain toll or toll-free facilities.
e) Public lands development roads and trails. The term "public lands development roads and trails" means those roads and trails that the Secretary of the Interior determines are of primary importance for the development, protection, administration, and utilization of public lands and resources under the control of the Secretary of the Interior.
f) Public lands highway. The term "public lands highway" means a forest road under the jurisdiction of and maintained by a public authority and open to public travel or any highway through unappropriated or unreserved public lands, nontaxable Indian lands, or other Federal reservations under thejurisdiction of and maintained by a public authority and open to public travel.
g) Public road. The term "public road" means any road or street under the jurisdiction of and maintained by a public authority and open to public travel.
h) Refuge road. The term “refuge road" means a public road that provides access to or within a unit of the National Wildlife Refuge System and for which title and maintenance responsibility is vested in the United States Government.

23 U.S.C. Sec. 120. Federal share payable

Emergency Relief - The Federal share payable on account of any repair or reconstruction provided for by funds made available under Section 125 of this title on account of any project on a Federal-aid highway, including the Interstate System, shall not exceed the Federal share payable on a project on such system as provided in subsections (a) and (b) of this section; except that (1) the Federal share payable for eligible emergency repairs to minimize damage, protect facilities, or restore essential traffic accomplished within 180 days after the actual occurrence of the natural disaster or catastrophic failure may amount to 100 percent of the costs thereof; and (2) the Federal share payable on account of any repair or reconstruction of forest highways, forest development roads and trails, park roads and trail, parkways, public lands highways, public lands development roads and trails, and Indian reservation roads may amount to 100 percent of the cost thereof.

The total cost of a project may not exceed the cost of repair or reconstruction of a comparable facility. As used in this section with respect to bridges and in Section 144 of this title, "a comparable facility" shall mean a facility which meets the current geometric and construction standards required for the types and volume of traffic which such facility will carry over its design life.

Sec. 125. Emergency relief

a) General Eligibility. Subject to this section and Section 120, an emergency fund is authorized for expenditure by the Secretary for the repair or reconstruction of highways, roads, and trails, in any part of the United States, including Indian reservations, that the Secretary finds have suffered serious damage as a result of:
   1) natural disaster over a wide area, such as by a flood, hurricane, tidal wave, earthquake, severe storm, or landslide; or
   2) catastrophic failure from any external cause.
b) Restriction on Eligibility. In no event shall funds be used pursuant to this section for the repair or reconstruction of bridges that have beenpermanently closed to all vehicular traffic by the State or responsible local official because of imminent danger of collapse due to a structural deficiency or physical deterioration.
c) Funding. Subject to the following limitations, there are authorized to be appropriated from the Highway Trust Fund (other than the Mass Transit Account) such sums as may be necessary to establish the fund authorized by this section and to replenish it on an annual basis:
   1) Not more than $100,000,000 is authorized to be obligated in any one fiscal year commencing after September 30, 1980, to carry out the provisions of this section; except that, if in any fiscal year the total of all obligations under this section is less than the amount authorized to be obligated in such fiscal year, the unobligated balance of such amount shall remain available until expended and shall be in addition to amounts otherwise available to carry out this section each year.
   2) Pending such appropriation or replenishment, the Secretary may obligate from any funds heretofore or hereafter appropriated for obligation in accordance with this title, including existing Federal-aid appropriations, such sums as may be necessary for the immediate prosecution of the work herein authorized. Funds

obligated under this paragraph shall be reimbursed from such appropriation or replenishment.

d) The Secretary may expend funds from the emergency fund herein authorized for the repair or reconstruction of highways on Federal-aid highways in accordance with the provisions of this chapter: Provided, that (1) obligations for projects under this section, including those on highways, roads, and trails mentioned in subsection (e) of this section, resulting from a single natural disaster or a single catastrophic failure in a State shall not exceed $100,000,000, and (2) the total obligations for projects under this section in any fiscal year in the Virgin Islands, Guam, American Samoa, and the Commonwealth of the Northern Mariana Islands shall not exceed $20,000,000.

   Notwithstanding any provision of this chapter, actual and necessary costs of maintenance and operation of ferryboats providing temporary substitute highway traffic service, less the amount of fares charged, may be expended from the emergency fund herein authorized on Federal-aid highways.

   Except as to highways, roads, and trails mentioned in subsection (e) of this section, no funds shall be so expended unless the Secretary has received an application therefore from the State highway department, and unless an emergency has been declared by the Governor of the State and concurred in by the Secretary, except that if the President has declared such emergency to be a major disaster for the purposes of the Robert T. Stafford Disaster Relief and Emergency Assistance Act (42 U.S.C. 5121 et seq.), concurrence of the Secretary is not required.

e) The Secretary may expend funds from the emergency fund herein authorized, either independently or in cooperation with any other branch of the Government, State agency, organization, or person, for the repair or reconstruction of forest highways, forest development roads and trails, park roads and trails, parkways, public lands highways, public lands development roads and trails, and Indian reservation roads, whether or not such highways, roads, or trails are on any of the Federal-aid highways.

f) Treatment of Territories. For the purposes of this section, the Virgin Islands, Guam, American Samoa, and the Commonwealth of Northern Mariana Islands shall be considered to be States and parts of the United States, and the chief executive officer of each territory shall be considered to be Governor of a State.

D.2. Title 23, Code of Federal Regulations

PART 668 EMERGENCY RELIEF PROGRAM

Subpart B Procedures for Federal Agencies for Federal Roads – Authority granted under: 23 U.S.C. 101, 120(e), 125 and 315; 49 CFR 1.48(b).

§ 668.201 Purpose.

To establish policy, procedures, and program guidance for the administration of emergency relief to Federal agencies for the repair or reconstruction of Federal roads which are found to have suffered serious damage by a natural disaster over a wide area or by catastrophic failure. [43 FR 59485, 12/21/1978]

§668.203 Definitions.

a) Applicant. Any Federal agency which submits an application for emergency relief and which has authority to repair or reconstruct Federal roads.
b) Betterments. Added protective features, such as, the relocation or rebuilding of roadways at a higher elevation or the extension, replacement or raising of bridges, and added facilities not existing priorto the natural disaster or catastrophic failure such as additional lanes, upgraded surfacing, or structures.
c) Catastrophic failure. The sudden failure of a major element or segment of a Federal road which is not primarily attributable to gradual and progressive deterioration or lack of proper maintenance. The closure of a facility because of imminent danger of collapse is not in itself a sudden failure.
d) Emergency repairs. Those repairs, including necessary preliminary engineering, construction engineering, and temporary traffic operations, undertaken during or immediately after a natural disaster or catastrophic failure (1) to restore essential travel, (2) to protect remaining facilities, or (3) to minimize the extent of damage.
e) Federal roads. Forest highways, forest development roads and trails, park roads and trails, parkways, public lands highways, public lands development roads and trails, and Indian reservation roads as defined under 23 U.S.C. 101(a).
f) Finding. A letter or other official correspondence issued by the Direct Federal Division Engineer (Direct Federal Division Engineer) to a Federal agency giving notification that pursuant to 23 U.S.C. 125, Federal roads have (Affirmative Finding) or have not (Negative Finding) been found to have suffered serious damage as the result of (1) a natural disaster over a wide area, or (2) a catastrophic failure.
g) Natural disaster. An unusual natural occurrence such as a flood, hurricane, severe storm, tidal wave, earthquake, or landslide which causes serious damage.
h) Permanent work. Repair or reconstruction to pre-disaster or other allowed geometric and construction standards and related Preliminary Engineering and Construction Engineering.
i) Direct Federal Division Engineer. Director of one of the Direct Federal field offices located in Vancouver, WA; Denver, CO; and Arlington, VA. [43 FR 59485, Dec. 12, 1978, as amended at 47 FR 10529, Mar. 11, 1982].

§ 668.205 Policy.

a) This emergency relief program is intended to pay the unusually heavy expenses in the repair and reconstruction of Federal roads resulting from damage caused by natural disasters over a wide area or catastrophic failures.
b) Emergency relief work shall be given prompt attention and priority over non-emergency work.
c) Permanent work shall be done by contract awarded by competitive bidding through formal advertising, where feasible.
d) It is in the public interest to perform emergency repairs immediately and prior approval or authorization from the Direct Federal Division Engineer is not required. Emergency repairs may be performed by the method of contracting (advertised contract, negotiated contract, or force account) which the applicant or the Federal

Highway Administration (where Federal Highway Administration performs the work) determines to be most suited for this work.

e) Emergency relief projects shall be promptly constructed. Projects not under construction by the end of the second fiscal year following the year in which the disaster occurred will be reevaluated by the Direct Federal Division Engineer and will be withdrawn from the approved Program of Projects unless suitable justification is provided by the applicant to warrant retention.
f) The Finding for natural disasters will be based on both the extraordinary character of the natural disturbance and the wide area of impact. Storms of unusual intensity occurring over a small area do not meet these conditions.
g) Diligent efforts shall be made to recover repair costs from the legally responsible parties to reduce the project costs where highway damages are caused by ships, barge tows, highway vehicles, vehicles with illegal loads, and similar improperly controlled objects or events.
h) Emergency funds shall not duplicate assistance under another Federal program or compensation from insurance or any other source. Where other funding compensates for only part of an eligible cost, emergency relief funding can be used to pay the remaining costs. [43 FR 59485, Dec. 12, 1978, as amended at 47 FR 10529, Mar. 11, 1982].

§ 668.207 Federal share payable from emergency fund.

The Federal share payable under this program is 100 percent of the cost. [43 FR 59485, Dec. 12, 1978]

§ 668.209 Eligibility of work.

a) Permanent work must have prior program approval in accordance with paragraph (a) of § 668.215 unless such work is performed as emergency repairs.
b) Emergency repairs, including permanent work performed incidental to emergency repairs, and all Preliminary Engineering may begin immediately and do not need prior program approval. Reimbursement shall be contingent upon the work ultimately being approved in accordance with the requirements of paragraph (a) of § 668.215.
c) To qualify for emergency relief, the damaged or destroyed road or trail must be designated as a Federal road.
d) Replacement highway facilities are appropriate when it is not practical and economically feasible to repair or restore a damaged element to its preexisting condition. Emergency relief is limited to the cost of a new facility constructed to current design standards of comparable capacity and character to the destroyed facility. With respect to a bridge, a comparable facility is one which meets current geometric and construction standards for the type and volume of traffic it will carry during its design life.
e) Emergency relief funds may participate to the extent of eligible repair costs when proposed projects contain betterments or other work not eligible for emergency funds.

f) Work may include:
   1) Repair to, or reconstruction of, seriously damaged highway elements for a distance which would be within normal highway right-of-way limits, including necessary clearance of debris and other deposits in drainage courses, where such work would not be classed as heavy maintenance.
   2) Restoration of stream channels when the work is necessary for the satisfactory operation of the federal road. The applicant must have responsibility and authority for maintenance and proper operation of stream channels restored.
   3) Betterments where clearly economically justified to prevent future recurring damage. Economic justification acceptable to the Direct Federal Division Engineer must weigh the cost of such betterments against the risk of eligible recurring damage and the cost of future repair.
   4) Actual Preliminary Engineering and Construction Engineering costs on approved projects.
   5) Emergency repairs. [43 FR 59485, Dec. 12, 1978, as amended at 47 FR 10529, Mar. 11, 1982].

§ 668.211 Notification, damage assessment, and finding.

a) Notification. During or as soon as possible after a natural disaster or catastrophic failure, each applicant will notify the Direct Federal Division Engineer of its tentative intent to apply for emergency relief and request that a Finding be made.
b) Acknowledgment. The Direct Federal Division Engineer will promptly acknowledge the notification and briefly describe subsequent damage assessment, Finding, and application procedures.
c) Field report. The applicant shall cooperate with the Direct Federal Division Engineer to promptly make a field survey of overall damage and in the preparation of a field report.
d) Finding. Using the field report and other information deemed appropriate, the Direct Federal Division Engineer will promptly issue a Finding, and if an Affirmative Finding is made, establish the date after which repair or reconstruction will be considered for emergency relief, and note the dates of the extraordinary natural occurrence or catastrophic event responsible for the damage or destruction.
e) Detailed site inspections.
   1) If an Affirmative Finding is made, the applicant shall cooperate with the Direct Federal Division Engineer to make a detailed inspection of each damage site.
   2) If it appears certain an Affirmative Finding will be made, the Direct Federal Division Engineer may elect to make these site inspections at the time damage is initially assessed pursuant to paragraph (c) of this section.
f) The Applicant shall make available to Federal Highway Administration personnel conducting damage survey and estimate work maps depicting designated Federal roads in the affected area. [43 FR 59485, Dec. 12, 1978, as amended at 47 FR 10529, Mar. 11, 1982].

§ 668.213 Application procedures.

a) Based on the detailed site inspections and damage estimates prepared pursuant to paragraph (e) of § 668.211, the Applicant will submit an application in the form of a letter to the Direct Federal Division Engineer which shall include a list of projects for which emergency relief is requested. The application shall be submitted within three (3) months after an Affirmative Finding.
b) The list of projects shall include emergency repairs, Preliminary Engineering, and permanent work, and provide for each project a location, length, project number, type of damage, description of work with a separate breakdown for betterments including a justification for those intended for emergency relief funding, proposed method of construction, estimated cost, and any other information requested by the Direct Federal Division Engineer.
c) If the initial list of projects is incomplete, a subsequent list(s) of projects shall be forwarded to the Direct Federal Division Engineer for approval consideration as soon as possible. [43 FR 59485, Dec. 12, 1978, as amended at 47 FR 10529, Mar. 11, 1982].

§ 668.215 Programming and project procedures.

a) The Direct Federal Division Engineer will advise the applicant in writing which projects in the application, or in any subsequent submittals pursuant to paragraph (c) of § 668.213 are approved, including any approval conditions. Approved projects shall constitute the approved Program of Projects (program).
b) Plans, specifications, and estimates shall be developed based on work identified in the approved program.
c) The Direct Federal Division Engineer will approve PS&Es, concur in the award of contracts or the rejection of bids, determine that construction by the force account method is in the public interest, and accept completed work in accordance with interagency procedures established by the Direct Federal Division Engineer.
d) The applicant shall notify the Direct Federal Division Engineer in writing of the semi-annual status and completion of each emergency relief project constructed by applicant forces. [43 FR 59485, Dec. 12, 1978, as amended at 47 FR 10529, Mar. 11, 1982].

APPENDIX E – DESIGNATED NATIONAL SCENIC AND HISTORIC TRAILS NATIONAL TRAILS SYSTEM ACT OF 1968

E.1. National Scenic Trails

a) Appalachian National Scenic Trails (NST)
b) Continental Divide NST
c) Florida NST
d) Ice Age NST

e) Natchez NST
f) North Country NST
g) Pacific Crest NST
h) Potomac Heritage NST

E.2. National Historic Trails

a) Ala Kahakai National Historic Trails (NHT)
b) California NHT
c) Captain John Smith Chesapeake NHT
d) El Camino Real de los Tejas NHT
e) El Camino Real de Tierra Adentro NHT
f) Iditarod NHT
g) Juan Bautista de Anza NHT
h) Lewis and Clark NHT
i) Mormon Pioneer NHT
j) Nez Perce (Nee-Me-Poo) NHT
k) Old Spanish NHT
l) Oregon NHT
m) Overmountain Victory NHT
n) Pony Express NHT
o) Santa Fe NHT
p) Selma to Montgomery NHT
q) Trail of Tears NHT

In: Emergency Relief Programs for U.S. Highways and Roads ISBN: 978-1-61942-858-4
Editors: Cody O. Harrison and Ted J. Gibson © 2012 Nova Science Publishers, Inc.

Chapter 3

HIGHWAY EMERGENCY RELIEF: STRENGTHENED OVERSIGHT OF PROJECT ELIGIBILITY DECISIONS NEEDED*

The United States Government Accountability Office

WHY GAO DID THIS STUDY

The Federal Highway Administration (FHWA), within the U.S. Department of Transportation (DOT), administers the Emergency Relief Program to provide funds to states to repair roads damaged by natural disasters and catastrophic failures. In 2007, GAO reported that in recent years states' annual demand for emergency relief funds often exceeded the program's $100 million annual authorization from the Highway Trust Fund and required supplemental appropriations from general revenues to address a backlog of funding requests from states. GAO recommended that FHWA tighten eligibility standards and coordinate with states to withdraw unneeded emergency relief funds, among other actions. For this report, GAO reviewed (1) Emergency Relief Program funding trends since 2007, (2) key program changes made in response to GAO's 2007 report, and (3) the extent to which selected emergency relief projects were approved in compliance with program eligibility requirements. GAO reviewed projects in New York, Texas, and Washington state, states selected based on the amount and frequency of funding allocations since 2007, among other factors.

WHAT GAO RECOMMENDS

GAO makes several recommendations including that FHWA establish (1) time frames to limit states' requests for emergency relief funds and to close completed projects and (2)

* This is an edited, reformatted and augmented version of The United States Government Accountability Office publication, Report to the Subcommittee on Transportation, Housing and Urban Development, and Related Agencies, Committee on Appropriations, House of Representatives, GAO-12-45, dated November 2011.

standardized procedures for reviewing emergency relief documentation and making eligibility decisions. DOT provided technical comments on project time frames and costs which GAO incorporated as appropriate.

WHAT GAO FOUND

From fiscal years 2007 through 2010, the Emergency Relief Program received about $2.3 billion, of which $1.9 billion came from three supplemental appropriations compared with about $400 million authorized from the Highway Trust Fund. FHWA allocated this funding to 42 states and 3 territories to reduce the backlog of funding requests, with $485 million in unfunded requests remaining as of June 2011. This backlog list did not include funding requests for August 2011 damages from Hurricane Irene. Because the program lacks time frames to limit states from requesting funds years after events occur, the June 2011 backlog list includes about $90 million for events that occurred prior to fiscal year 1994. Without time limits for emergency relief funding requests, FHWA's ability to anticipate and manage future program costs is hindered.

In response to GAO's 2007 report, FHWA withdrew about $367 million of unobligated emergency relief funds from states and redistributed most of this funding for other emergency relief needs. However, additional funding remains unused, including (1) at least $63 million allocated to states before fiscal year 2007 that has yet to be obligated to projects and (2) $341 million obligated between fiscal years 2001 and 2006 that remains unexpended. Due to a lack of time frames for states to close-out completed projects, FHWA lacks project status information to determine whether unexpended funding is no longer needed and could be deobligated. FHWA has not addressed GAO's 2007 recommendation to revise its regulations to limit the use of emergency relief to fully fund projects that have grown in scope and cost as a result of environmental or community concerns. The Emergency Relief Program faces the continued risk of escalating costs due to projects that have grown in scope beyond the program's goal of restoring damaged facilities to predisaster conditions.

Instances of Missing or Incomplete Emergency Relief Project Documentation

GAO-identified areas of concern regarding eligibility	Instances
Missing or incomplete detailed damage inspection reports	47 of 83
Missing repair cost estimates	42 of 83
Missing or incomplete dates for 100 percent federal funding projects	39 of 58
Missing documentation for specific improvements	6 of 15

Source: GAO analysis of Emergency Relief Project documentation in three FHWA state offices.

GAO's review of 83 emergency relief project files in three FHWA state offices found many instances of missing or incomplete documentation—as such, GAO was unable to determine the basis by which FHWA made many eligibility determinations. For example, about half of the project files did not include required repair cost estimates, and 39 of 58 (67 percent) emergency repair projects approved for 100 percent federal funding did not contain documentation of completion within 180 days—a requirement for states to receive 100 percent federal funding. FHWA lacks clear requirements for how states submit and FHWA approves key project documentation, which has resulted in FHWA state offices applying eligibility guidelines differently. Establishing standardized procedures for reviewing emergency relief documentation and making eligibility decisions would provide greater assurance that projects are in fact eligible and that FHWA makes eligibility determinations consistently and transparently.

ABBREVIATIONS

Caltrans	California Department of Transportation
DDIR	detailed damage inspection report
DOT	U.S. Department of Transportation
FHWA	Federal Highway Administration
FMIS	Fiscal Management Information System
SAFETEA-LU	Safe, Accountable, Flexible, Efficient Transportation Equity Act: A Legacy for Users
S.R.1	State Route 1
TXDOT	Texas Department of Transportation
WSDOT	Washington State Department of Transportation

November 8, 2011

The Honorable Tom Latham
Chairman

The Honorable John W. Olver
Ranking Member
Subcommittee on Transportation, Housing and
Urban Development, and Related Agencies
Committee on Appropriations
House of Representatives

In recent years, many states have experienced natural disasters, such as hurricanes, floods, and storms, which have caused catastrophic damage to transportation infrastructure and overwhelmed the capacity of state and local governments to respond and recover. Reconstruction after these events can cost taxpayers billions of dollars. As part of the continuing federal role in responding to and recovering from natural disasters and similar events, the Federal Highway Administration (FHWA), within the U.S. Department of Transportation (DOT), administers the Emergency Relief Program which provides funding to repair or reconstruct federal-aid highways and roads on federal lands damaged or destroyed by natural disasters and other catastrophic events.[1]

Since 1972, Congress has authorized $100 million annually in contract authority for FHWA's Emergency Relief Program to be paid from the Highway Trust Fund.[2] However, in recent years the Highway Trust Fund has not been the only source of funds for the Emergency Relief Program. The $100 million annual authorization of contract authority has remained constant since 1972, and due to inflation, it has declined in real value. States' need for assistance from fiscal years 1998 through 2006 consistently exceeded the $100 million annual authorization of contract authority, resulting in a backlog of funding requests that reached $740 million in 2004. As a result, the program has relied on supplemental appropriations for 86 percent of its funding from fiscal years 1998 through 2006.[3] In past years, supplemental appropriations were drawn from the Highway Trust Fund. However, with the enactment of the Safe, Accountable, Flexible, Efficient Transportation Equity Act: A Legacy for Users (SAFETEA-LU) in August 2005, Congress authorized additional necessary funding for the Emergency Relief Program in excess of the $100 million annual contract authority to be appropriated from general revenues.[4]

In our 2007 report, we identified a significant fiscal imbalance between available funds and eligible projects in the Emergency Relief Program.[5] We found that FHWA was not recapturing or redistributing unused Emergency Relief Program allocations to states with immediate program needs, as specified in FHWA's program guidance. We also raised concerns about the use of emergency relief funds to fully finance projects whose scope and costs had grown as a result of environmental and community concerns. We made several recommendations to FHWA to improve its management, including ensuring that unneeded emergency relief allocations were withdrawn on a timely basis so the backlog of unfunded requests could be addressed, and to revise program regulations to tighten the eligibility criteria for emergency relief funding, among other recommendations.

As requested and in light of these concerns, this report reviews (1) FHWA Emergency Relief Program funding trends since our 2007 report, (2) key changes to the Emergency Relief Program implemented in response to concerns raised in our 2007 report, and (3) the extent to which selected emergency relief projects were approved in compliance with program eligibility requirements.

To identify Emergency Relief Program funding trends and key changes made to the program since our 2007 report, we reviewed federal statutes, including supplemental appropriation acts, and FHWA data on emergency relief allocations to states from fiscal years 2007 through 2010. Our scope of work did not include any Emergency Relief Program activities in response to Hurricane Irene, which occurred in August 2011. We also reviewed and analyzed financial data from FHWA's fiscal management information system (FMIS) on emergency relief allocations and obligations to states as of May 31, 2011, as well as funds that were obligated to and expended by states for events occurring from fiscal years 2001 through 2010. We reviewed the procedures used by FHWA to enter and verify data into FMIS, and we found the data to be sufficiently reliable for our purposes. We also reviewed FHWA Emergency Relief Program regulations and guidance, including FHWA's *Emergency Relief Manual* as revised in 2009. We interviewed FHWA officials in the Office of Program Administration to determine why and how specific changes were made to the program.

To determine the extent to which selected emergency relief projects were approved in compliance with program eligibility requirements, we reviewed federal statutes, regulations, and FHWA guidance on emergency relief eligibility requirements and examined a sample of emergency relief project files in three FHWA division offices. We selected a nongeneralizable sample of 88 emergency relief project files from FHWA division offices in three states—New York, Texas, and Washington state—to demonstrate the range of practices and projects that the Emergency Relief Program funds across the country. Five of the 88 projects in our review had been withdrawn by states because, in part, FHWA had determined them ineligible for emergency relief funds, bringing the total number of projects reviewed to

83. We selected New York, Texas, and Washington state because they were among the states receiving the most funding allocations from fiscal years 2007 through 2010, among other factors. We chose the files according to several criteria, including a criterion to examine a mix of active and closed projects that were obligated more than $1 million in emergency relief funds. Prior to our site visits, we requested that the division offices provide all documentation they maintain for each of the projects selected in our sample, which represented approximately 67 percent of all emergency relief funds that FHWA obligated to those states during that time period. We reviewed all the documentation provided during our site visits, and requested follow-up information as necessary. To gather additional information on the project files, we reviewed the procedures used to manage and oversee emergency relief projects and interviewed officials in the FHWA division offices and state departments of transportation for all three states. We provided the results of our file review to FHWA for their comment and incorporated their responses as appropriate.

We conducted this performance audit from November 2010 to November 2011 in accordance with generally accepted government auditing standards. Those standards require that we plan and perform the audit to obtain sufficient, appropriate evidence to provide a reasonable basis for our findings and conclusions based on our audit objectives. We believe that the evidence obtained provides a reasonable basis for our findings and conclusions based on our audit objectives. Appendix I provides a more detailed description of our scope and methodology.

We provided a copy of this report to DOT for review and comment. DOT officials provided technical comments by e-mail, including information on Emergency Relief Program time frames and growth of project costs. We incorporated this information into the draft as appropriate.

BACKGROUND

The Emergency Relief Program, authorized by section 125 of title 23 of the U.S. Code, provides assistance to repair or reconstruct federal-aid highways and roads on federal lands that have sustained serious damage from natural disasters or catastrophic failures. Congress has provided funds for this purpose since at least 1938. Examples of natural disasters include floods, hurricanes, earthquakes, tornadoes, tsunamis, severe storms, and landslides. Catastrophic failures qualify if they result from an external cause that leads to the sudden and complete failure of a major element or segment of the highway system that has a disastrous impact on transportation. Examples of qualifying causes of catastrophic failures include acts of terrorism or incidents such as a barge striking a bridge pier causing the sudden collapse of the structure or a truck crash resulting in a fire that damages the roadway. For natural disasters or other events to be eligible for emergency relief funding, the President must declare the event to be an "emergency" or a "major disaster" under the Robert T. Stafford Disaster Relief and Emergency Assistance Act[6] or the governor must declare an emergency with the concurrence of the Secretary of Transportation.[7]

Since 1972, Congress has authorized $100 million annually in contract authority for the Emergency Relief Program to be paid from the Highway Trust Fund. Accordingly, FHWA may obligate up to $100 million in any one fiscal year for the program. Any unobligated balance remains available until expended.[8]

Additionally, obligations to a single state resulting from a single natural disaster or a single catastrophic failure may not exceed $100 million.[9] In some cases, Congress has enacted legislation lifting this cap for large-scale disasters. Moreover, as provided in FHWA's regulations, states are eligible for assistance under the Emergency Relief Program if the cost of the damage from a single event exceeds $700,000 for emergency assistance.[10] The $700,000 threshold includes the damage cost for all sites in any state affected by the disaster. According to FHWA guidance, each prospective damage site must have at least $5,000 of repair costs to qualify for funding assistance—a threshold intended to distinguish unusually large expenses eligible for emergency relief funding from costs that should be covered by normal state maintenance funding.

The Emergency Relief Program's authorizing statute and FHWA's regulations and guidelines distinguish between emergency and permanent repairs. Emergency repairs are made during and immediately following a disaster to quickly restore essential highway traffic service, minimize the extent of damage, or protect remaining facilities, including removing debris and constructing detours and temporary roadway surfaces. Permanent repairs are undertaken, normally after emergency repairs have been completed, to restore seriously damaged highway facilities to predisaster conditions. In some instances, such as the destruction of a bridge, complete replacement may be needed. In these cases, the bridge would be rebuilt to current design standards[11] (see fig. 1 for photographs of typical emergency and permanent repair examples).

The Emergency Relief Program may fund up to 100 percent of emergency repair project costs incurred within the first 180 days following an eligible disaster.[12] The program funds permanent repair projects and emergency repair project costs after the first 180 days at the percentage normally provided for work on that type of federal-aid highway.[13] For example, the federal share for interstate highway projects is 90 percent of the cost, and the federal share for most other federal projects is 80 percent.[14] Also, construction on permanent repairs must begin by the end of the second fiscal year following the year in which the disaster occurred;[15] however, FHWA may grant time extensions for projects needing extensive environmental evaluation, litigation, or complex right-of-way.

Sources: FHWA Washington state and Texas division offices.

Emergency repair example: debris removal

Permanent repair example: bridge replacement

Figure 1. Examples of Emergency and Permanent Repairs.

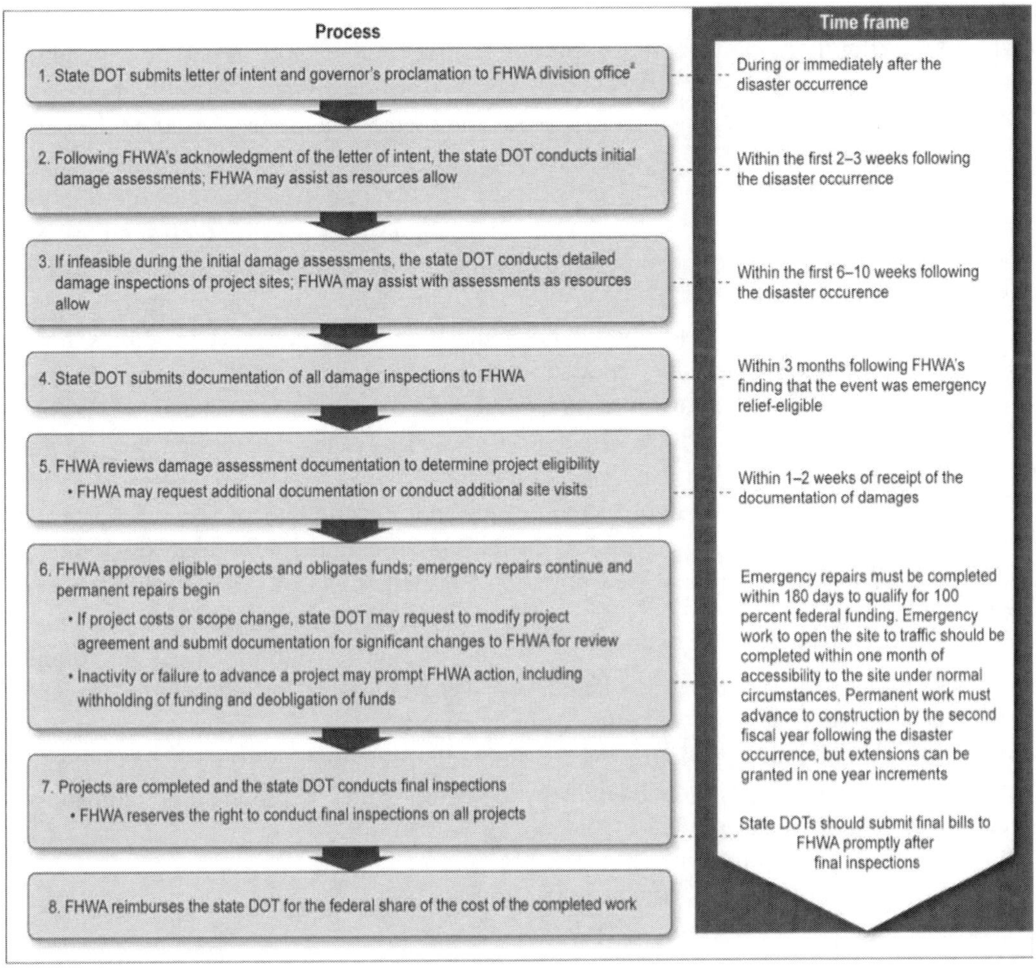

Source: GAO analysis of FHWA information.

[a] A governor's declaration is not required if the President has declared the event to have been an emergency or major disaster under the Robert T. Stafford Disaster Relief and Emergency Assistance Act.

Figure 2. Emergency Relief Project Processes and Areas of Responsibility.

In addition, the program is not intended to pay for "betterments," projects that change the function or character of the highway facility, such as expanding road capacity. However, FHWA may determine that betterments are eligible for program funding if they pass a benefit-cost test that weighs their cost against the prospective cost to the Emergency Relief Program for potentially chronic future repairs.

Figure 2 shows processes and areas of responsibilities for the Emergency Relief Program.

FHWA division offices located in each state review applications for emergency relief submitted by the state departments of transportation and decide the eligibility of emergency relief projects. Once the division office determines that an event is eligible for funding, it requests an emergency relief allocation from FHWA's Office of Program Administration, which manages the allocation of emergency relief to division offices.[16] If funding is not immediately available, the request is added to a nationwide list of funding requests, known as the emergency relief backlog list, which is used to apprise Congress of states' requests for funding for specific events. As funding becomes available for obligation—either through the program's $100 million annual authorization of contract authority or through a supplemental appropriation from Congress—FHWA enters into a project agreement for individual projects formalizing the conditions of its project approval and incurs obligations, based on the damage estimates prepared by states and approved by FHWA.[17]

As with other federal-aid highway programs, FHWA reimburses the states for emergency relief work as that work is completed and invoiced. If emergency relief funding is not immediately available for emergency relief work, FHWA may obligate other available transportation funds[18] as may be necessary for the immediate execution of emergency relief work and reimburse any funds actually expended once emergency relief funds become available.[19] Because estimates are based on circumstances existing at the time funds are obligated, more funding is sometimes obligated than is ultimately needed. FHWA may deobligate unexpended funds which will not be needed by states, increasing the amount of funds available for obligation for use for other projects.

The level of FHWA's oversight in the design and construction of federal-aid projects, including emergency relief, is determined by the classification of the roadway. Since the passage of the Intermodal Surface Transportation Efficiency Act in 1991,[20] states have been allowed to assume an increased amount of oversight responsibility for the design and construction of federal-aid projects. However, this expanded authority does not diminish FHWA's responsibility to determine whether projects are eligible for federal funds or ensure that federal funds are efficiently and effectively managed. States are primarily responsible for oversight of federal-aid highway projects, including emergency relief, on federal-aid highways that are off the National Highway System—which represent approximately 83 percent of the nation's road miles.

Table 1. Types of Projects Receiving State Oversight versus FHWA Oversight

Type of project	Mileage	Design and construction oversight	Exceptions
Federal-aid highways off the National Highway System	798,000	State assumes oversight responsibilities for design, plans, specifications, estimates, contract awards, and inspection of projects.	State determines state oversight is not appropriate
National Highway System, non-Interstate routes	115,000	State may assume the oversight responsibilities for design, plans, specifications, estimates, contract awards, and inspections of projects.	State or FHWA determines state oversight is not appropriate
National Highway System, Interstate routes	47,000	FHWA exercises full oversight responsibilities to (1) prescribe design and construction standards, (2) approve design plans and estimates, (3) approve the selection of the contract award, (4) inspect the progress of construction, and (5) render final acceptance on projects when they are completed.	Certain types of projects, or projects below a dollar threshold, where FHWA and state have agreed state oversight is appropriate

Source: GAO analysis.

In addition, states may assume responsibility for National Highway System projects not on the Interstate Highway System.[21] State departments of transportation may approve design plans and estimates and conduct project inspections to ensure the completion of emergency relief projects, among other things. FHWA exercises full oversight responsibility for many projects on the Interstate Highway System, as shown in table 1.[22]

Although FHWA administers the Emergency Relief Program in the same manner as the regular federal-aid highway program, there are important differences between the two.

The regular program derives revenues from highway users and funding is provided on a multiyear basis through formulas designed to ensure that each state receives its "fair share" based on estimated contributions to the Highway Trust Fund. Since the passage of SAFETEA-LU, a majority of emergency relief funds, as we have reported, have derived from general revenues, for expenses that for the most part cannot be planned for, and states compete for funding based on need. While formula funding for the regular federal-aid highway program is limited by amounts annually apportioned to the states, any emergency relief funds that a state receives are in addition to such apportionments.

And while regular federal-aid program projects require a state to match its federal funding amount, emergency repair projects may qualify for 100 percent federal funding.

Our 2007 report identified concerns regarding FHWA's financial oversight of the Emergency Relief Program and the program's eligibility criteria and made several recommendations to FHWA.[23]

- First, we found that FHWA was not routinely recapturing all unused program funds allocated to states—including unused unobligated emergency relief funds as well as unexpended obligated funds—that were not needed for eligible projects. This contributed to a situation in which states with immediate disaster needs were placed on the backlog list of states requesting funds, while states with no current disaster needs retained their allocations.[24] We therefore recommended that FHWA division offices annually coordinate with states to identify unused emergency relief funds and withdraw any unneeded amounts. Funds in excess of any outstanding need, if any, would be identified to Congress for rescission or to reduce future appropriations.
- Second, we expressed concern that the scope of eligible activities had expanded in recent years as a result of congressional waivers or FHWA rulemaking to revise eligibility criteria. As a result, emergency relief funds were being used to fully finance projects that had expanded in scope and costs beyond the typical emergency relief project. To address this concern, we recommended that FHWA tighten the eligibility criteria for funding, which could include limitations on the use of emergency relief funds to fully finance projects that grew in scope and costs as a result of environmental and community concerns.
- Third, we noted that the lack of a standard definition of what constitutes an eligible damage site might allow many smaller costs to be charged against the program and recommended that FHWA clarify its program manual to better define an eligible site.

SUPPLEMENTAL APPROPRIATIONS COMPRISE MOST EMERGENCY RELIEF FUNDING PROVIDED TO STATES, AND A BACKLOG OF FUNDING REQUESTS REMAINS

From Fiscal Years 2007 through 2010, Congress Provided More than $2.3 Billion for Emergency Relief Events and to Address a Backlog of Unfunded Requests

From fiscal years 2007 through 2010, Congress provided more than $2.3 billion to the Emergency Relief Program, including more than $1.9 billion in three supplemental appropriations from general revenues and about $400 million in contract authority paid from the Highway Trust Fund (see fig. 3). The supplemental appropriations represented 83 percent of the program's funding over that time period. This percentage has been fairly consistent over time: 86 percent of the total Emergency Relief Program funding provided from fiscal years 1990 through 2006 came from supplemental appropriations.[25]

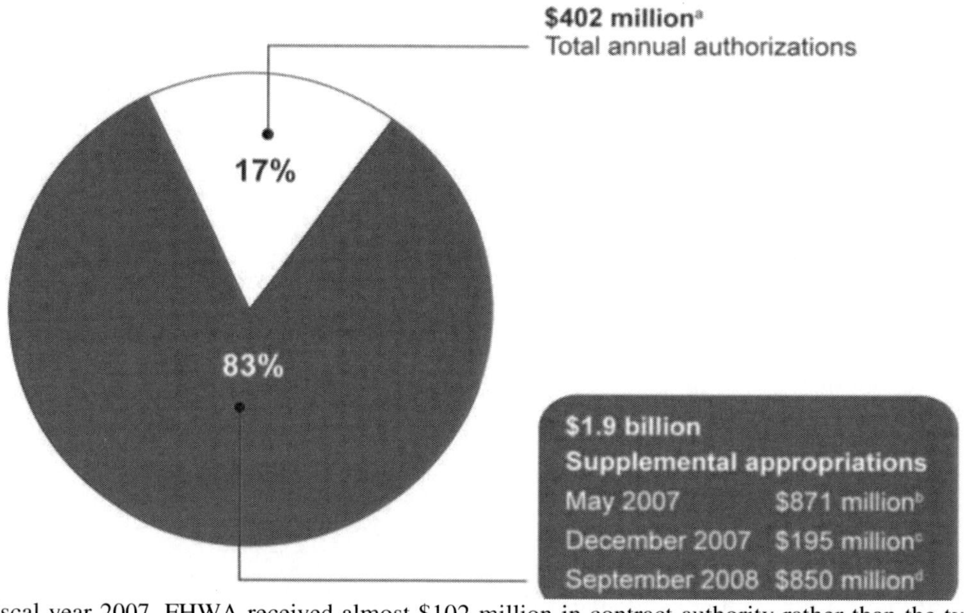

[a]In fiscal year 2007, FHWA received almost $102 million in contract authority rather than the typical $100 million. The additional $2 million was the Emergency Relief Program's share of additional budget authority provided to all highway programs to align total highway budget authority with revised revenue from the Highway Trust Fund for that year.

[b] U.S. Troop Readiness, Veterans' Care, Katrina Recovery, and Iraq Accountability Appropriations Act, 2007, Pub. L. No. 110-28, title IV, ch. 8, 121 Stat. 112, 160 (2007).

[c] Consolidated Appropriations Act, 2008, Pub. L. No. 110-161, div. K, title I, 121 Stat. 1844, 2384 (2008).

[d] Consolidated Security, Disaster Assistance, and Continuing Appropriations Act, 2009, Pub. L. No. 110-329, div. B, title I, ch. 10, 122 Stat. 3574, 3598 (2008).

Figure 3. Total Emergency Relief Program Funding, Fiscal Years 2007 through 2010.

Two of the supplemental appropriations that Congress provided to the Emergency Relief Program since fiscal year 2007 were used to address the backlog of unfunded emergency relief requests from states. In May 2007, Congress provided $871 million[26] to help clear a backlog of $736 million in funding requests from 46 states. In September 2008, when the backlog list reached more than $560 million, Congress provided $850 million to address this backlog and provide additional funds for future requests.[27] In December 2007, Congress provided $195 million for the reconstruction of the Interstate 35 West Bridge in Minnesota.[28]

FHWA has allocated all of the $2.3 billion provided to the program since fiscal year 2007, as well as an additional $100 million carried over from previously provided program funding, among 42 states and three territories.[29] Sixty-five percent of the allocations (almost $1.6 billion) went to six states—California, Louisiana, Minnesota, North Dakota, Texas, and Washington state (see fig. 4). California received almost $538 million, the most of all states, and most of this was a result of the 2005–2006 winter storms. Washington state was allocated almost $166 million in response to 10 events ranging from a single event estimated to cost $1 million to about $58 million to respond to flooding caused by severe rains in December 2007.

In our 2007 report we noted that Congress has, on occasion, expanded Emergency Relief Program eligibility criteria and increased the amount of program funding to respond to specific disasters.[30] For example we noted that Congress directed the Emergency Relief Program to fund 100 percent of all repair and reconstruction of highways, roads, and bridges necessitated by Hurricanes Katrina, Rita, and Wilma, because the states' resources were inadequate to deal with the string of disasters.[31] From fiscal years 2007 through 2010, FHWA allocated about $484 million in emergency relief in response to two events—one in Minnesota and one in North Dakota.

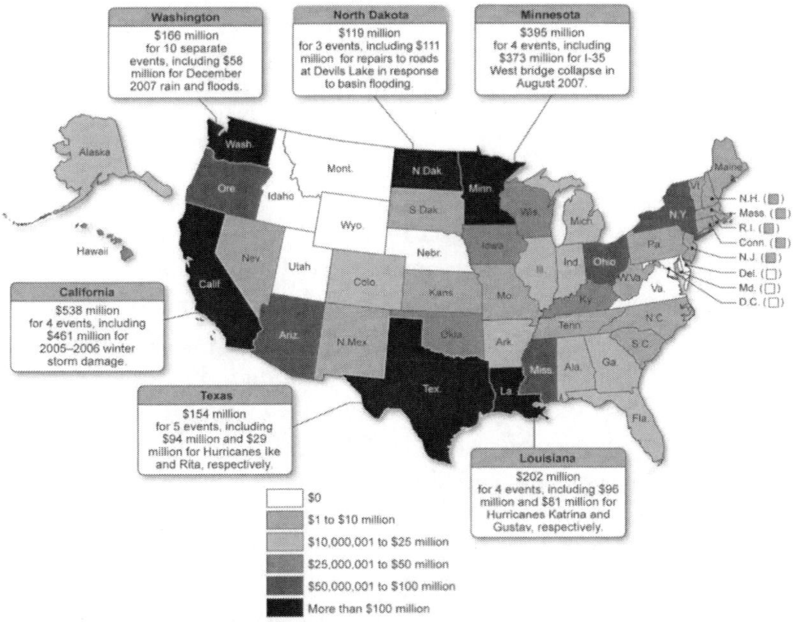

Source: GAO analysis of HWA data: Map Rescurces (map).

Note: The three territories that were provided allocations from fiscal years 2007 through 2010 were American Samoa ($23 million), Puerto Rico ($20 million), and U.S. Virgin Islands ($730,591).

Figure 4. Emergency Relief Allocations by State, Fiscal Years 2007 through 2010.

- *Collapse of the Interstate 35 West Bridge in Minneapolis, Minnesota.* Since 2007, FHWA allocated about $373 million [32] for the reconstruction and repair of an 8-lane, 1,900 foot bridge spanning the Mississippi River which collapsed on August 1, 2007, killing 13 people and injuring 145 others.[33] Five days after the disaster, Congress passed legislation that authorized the use of emergency relief funds for the repair and reconstruction of the bridge, waived the $100 million limitation on emergency relief obligations, and established a 100 percent federal share for all repair costs for the project.[34]
- *Basin flooding of roadways at Devils Lake, North Dakota.* Since fiscal year 2007, North Dakota was allocated more than $111 million in emergency relief funding for projects associated with Devils Lake—a large natural basin that lacks an outlet for rising water to flow out of the lake. Starting in the early 1990s, the lake level has risen dramatically, threatening adjacent roadways. Although Emergency Relief Program regulations define a natural disaster as a sudden and unusual natural occurrence, FHWA determined that the gradual and predictable basin flooding at Devils Lake is eligible for Emergency Relief Program funding.35 In 2005, through SAFETEA-LU, Congress authorized up to $10 million of Emergency Relief Program funds to be expended annually, up to a total of $70 million, to address an additional problem at Devils Lake and make repairs to certain roads which were impounding water and acting as dams.36 In the absence of other authority, this funding must come out of the $100 million annual authorization of contract authority, effectively reducing the annual emergency relief funding available to other states. As of March 2010, the Emergency Relief Program has provided more than $256 million for projects related to Devils Lake flooding.37

Emergency Relief Faces Risk from Escalating Costs of Events Occurring in Past Years

In recent years, Congress has provided significant supplemental funding to the Emergency Relief Program, but as of June 2011, a $485 million backlog of funding requests from states remained. This backlog did not include funding requests for August 2011 damages from Hurricane Irene. The backlog list provides a snapshot of states' funding requests at a given time and is subject to change as states experience new eligible events. According to guidance in FHWA's *Emergency Relief Manual*, requested amounts are based on the states' anticipated need for emergency relief for the current fiscal year and may be less than the total emergency relief needs for any specific event.

The June 2011 backlog list contained almost $90 million in formal funding requests for several events that occurred between 1983 and 1993 that were previously determined to be eligible by FHWA. Specifically, California requested almost $83 million for a single, long-term project in response to a 1983 rockslide, known as Devil's Slide, and an additional $6.5 million for four other events from fiscal years 1990 through 1993.

According to FHWA, these requests are for approved emergency relief events with projects that have had delays due to environmental issues or cost overruns.

Once an event has been approved for emergency relief by FHWA, current program rules do not establish a time limit in which states must submit all funding requests for repairs. Although FHWA requires states to submit a list of projects within three months of approving a state's application for emergency relief,[38] eligibility stemming from an approved event does not lapse, and a state's list of projects may be amended at any time to add new work. Consequently, FHWA faces the risk of receiving reimbursement requests from states for projects years after an event occurs, including requests for projects that have experienced significant delays and cost increases over time, due to environmental or community concerns. The June 2011 backlog list included project funding requests for two events that occurred more than 10 years ago and which demonstrate FHWA's risk of escalating long-term costs due to older events.

Devil's Slide tunnel for State Route 1 in California. Total costs estimated to be more than $631 million in emergency relief funds resulted from the ongoing construction of a tunnel in response to a 1983 landslide in California. The Devil's Slide area in California is an unstable cliff formation on the Pacific coast near San Francisco that is subject to reoccurring rock slides onto State Route 1 (S.R.1). Following a major landslide over the winter of 1983 that closed S.R.1 for 3 months, the California Department of Transportation (Caltrans) began to pursue the idea of relocating S.R.1 away from the slide area. The project required a comprehensive environmental impact study, which took several years to complete. The study was subsequently challenged in U.S. District Court, which delayed construction for several more years and resulted in an additional environmental review. As we reported in 2007, in the years that passed since the original environmental impact study, community attitudes shifted in favor of relocating S.R.1 by way of a tunnel through the mountain behind Devil's Slide so that S.R.1 would not be affected by future rock slides. In 1998, Congress declared the Devil's Slide project to be eligible for emergency relief[39] and, after an additional environmental review, the tunnel alternative was selected in 2002 and construction of a pair of 4,200-foot-long, 30-foot-wide tunnels began in 2006—23 years after the originating emergency relief event.

Construction of the tunnel is ongoing, with a planned completion in March 2013. To date, FHWA has obligated about $555 million in emergency relief funds to the Devil's Slide tunnel project out of an estimated cost of $631 million. The $631 million total project cost estimate includes the $83 million requested on the June 2011 backlog list—which is for work completed during fiscal year 2011—as well as an additional $120 million to be requested in the future to fully reimburse Caltrans to complete the project.

Alaskan Way Viaduct in Seattle, Washington. The June 2011 backlog list also contained a "pending" request of $40.5 million from Washington state in response to a February 2001 earthquake which damaged the Alaskan Way Viaduct—a 2-mile double-deck highway running along Seattle's waterfront.[40] In the months after the event, FHWA approved $3.6 million for emergency relief repairs to cracks in several piers supporting a section of the viaduct, which were completed by December 2004. At the time of the earthquake, the Washington State Department of Transportation (WSDOT) had begun considering options for replacing the viaduct, which was approaching the end of its design life. After continued monitoring, WSDOT found that the viaduct had experienced accelerated deterioration as a result of the earthquake and requested $2 billion in emergency relief to replace the viaduct. Congress directed FHWA and state and local agencies to determine the specific damages caused by the earthquake and the amount eligible for emergency relief.[41] In response, FHWA found that while the replacement of the entire viaduct was not eligible for emergency relief, the project was eligible to receive $45 million to replace the section of the viaduct damaged by the earthquake.[42] FHWA further found that if WSDOT decided to move forward with a more comprehensive replacement project for the entire facility, the estimated amount of emergency relief eligibility could be applied to that project. WSDOT now plans to replace the entire viaduct with a bored tunnel under downtown Seattle, with an estimated cost of almost $2 billion. According to FHWA's Washington state division office, the $40.5 million listed on the June 2011 emergency relief backlog list will be obligated toward the construction of the larger replacement project for the viaduct.

The lack of a time limit for states to submit emergency relief funding requests raises the risk of states filing claims for additional funding years after an event's occurrence, particularly for projects that grow significantly in cost or scope over time. States may have good reasons for submitting funding requests years after an event—particularly for larger-scale permanent repairs that may take years to complete—but such projects can grow unpredictably. The example of the relocation of S.R.1 away from Devil's Slide and the cost and scope increases that resulted from more than two decades of delays to complete lengthy environmental reviews and address community concerns is case and point. The absence of a time limit for states to submit funding requests hinders FHWA's ability to manage future claims to the program and creates a situation where Congress may be asked to provide additional supplemental appropriations for emergency relief years after an event occurs. Furthermore, states requesting emergency relief funds for projects many years after an event raises questions as to whether the repairs involved meet the goal of the Emergency Relief Program to restore damaged facilities to predisaster conditions.

In 2007 we recommended that FHWA revise its regulations to tighten program eligibility criteria, which could include limitations on the use of emergency relief funds to fully finance projects that grew in scope and cost as a result of environmental and community concerns. In July 2011, DOT's regulatory agenda43 included a planned rulemaking for the Emergency Relief Program that would, among other actions, consider specific time restrictions for states when filing a claim for emergency relief eligible work.44 However, in October 2011, FHWA withdrew this planned item from its agenda. According to an FHWA official, the planned rulemaking was withdrawn because it was premature and because FHWA is still determining what changes if any are needed to address GAO's 2007 recommendations.

FHWA's Program Revisions Have Not Fully Addressed Prior Concerns

FHWA Now Has Procedures to Withdraw Some Unused Emergency Relief Allocations from States, But Lacks Information to Verify Whether Additional Unused Allocations Are Still Needed

Since our 2007 report, FHWA has implemented a process to withdraw unused allocations and reallocate funding to benefit other states. FHWA undertook these actions in response to our recommendation to require division offices to annually coordinate with states to identify and withdraw unused allocations that are no longer needed so funds may be used to reduce the backlog of other program requests.[45] Since 2007, FHWA has based its allocations on a state's estimate of anticipated emergency relief obligations for the fiscal year. Prior to fiscal year 2007, FHWA's policy was to allocate the full amount of each state's emergency relief request, based on total available program funds.

In fiscal years 2010 and 2011, FHWA division offices coordinated with states to identify and withdraw unused allocations representing approximately $367 million in emergency relief funds from a total of 25 states and 2 territories.[46] To withdraw unused funds from states, FHWA reviews its financial database, FMIS, to identify the amount allocated to each state that has not been obligated to specific projects. FHWA then asks each state to identify remaining fiscal year need for new obligations and the amount of any allocations that will no longer be needed. FHWA then withdraws the amount determined by the state to be no longer needed and reallocates that amount to other nationwide emergency relief needs, such as unfunded requests on the backlog list. Most of the withdrawn allocations were originally allocated to states from fiscal years 2003 to 2006, as shown in figure 5. Of the $299 million that was withdrawn for events occurring from fiscal years 2003 to 2006, about $230 million was withdrawn from Florida.

FHWA reallocated $295 million of the $367 million withdrawn from states for other nationwide requests.[47] According to FHWA, the remaining $72 million that was withdrawn but not yet reallocated will be made available to states in future allocations.

As of the end May 2011, $493 million that FHWA allocated to states in response to events occurring since 1989 remains unobligated. [48] A significant portion of this amount likely reflects the recent allocation of $320 million in April 2011. However, at least $63 million of the unobligated balance is for older allocations, provided prior to fiscal year 2007.

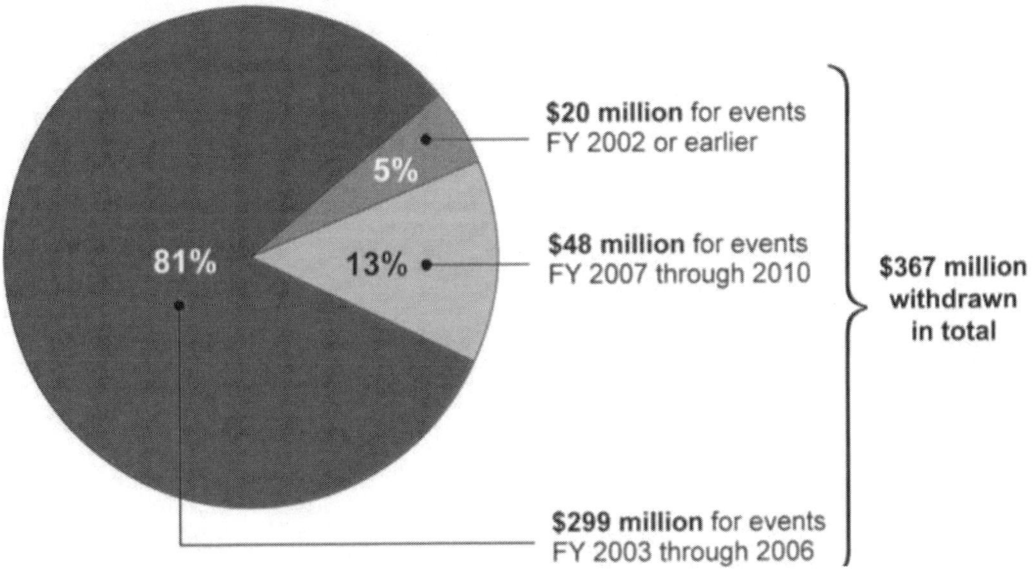

Source: GAO analysis of HWA data.

Note: Percentages do not add to 100 percent due to rounding.

Figure 5. Unobligated Emergency Relief Allocations Withdrawn from States since 2007 by the Fiscal Year in Which the Events Occurred.

Specifically, New York's unobligated balance includes almost $52 million provided after the September 11, 2001, terrorist attacks for roadway repairs delayed due to ongoing building construction around the World Trade Center site.[49] FHWA's New York division reported that these repairs are not expected to be completed until 2014. In addition, California maintained an unobligated balance of more than $11 million from the October 1989 Loma Prieta earthquake. According to FHWA California division officials, FHWA sought to withdraw some of this allocation, but Caltrans and local officials indicated that this allocation was necessary to complete environmental mitigation and bike path projects that were part of reconstruction of the collapsed Bay Bridge connecting San Francisco and Oakland in California.

Although the *Emergency Relief Manual* states that FHWA division offices are to identify and withdraw unused program funding allocations annually, we found several instances in which division offices applied unused allocations from existing events to new events in the same state without requesting a new allocation.

Specifically, our file review at the FHWA Washington state and New York state division offices identified three events from fiscal years 2009 and 2010 that the division offices approved as eligible and funded with allocations that were no longer needed from previous events. This practice, which was permitted in the 1989 version of the *Emergency Relief Manual*, limits FHWA's ability to track unobligated balances for specific events and determine whether those funds are no longer needed and may be withdrawn.

FHWA took steps to limit divisions from using this practice by removing language permitting the practice in the 2009 *Emergency Relief Manual*. According to FHWA, this change was made so that funds could be more equitably distributed across the nation to address the backlog of funding requests, rather than allowing states to hold unused funds in reserve for future events.

Although FHWA removed the language permitting this practice from the manual, FHWA has not provided written guidance to its divisions to prohibit them from applying unused allocations to new events in the same state, and the practice is still being used. For example, in February 2011, FHWA's headquarters allowed the Washington state division to shift unused funds from a prior event to a new event, and in doing so, the division office did not submit a request for an allocation of funds for those new events and FHWA headquarters did not provide an allocation for those events. Consequently, FHWA headquarters did not have a record for the events, nor did it know the amount of funds made available by the division for these events. Furthermore, FHWA headquarters officials were unable to determine how prevalent this practice was across division offices. As a result, FHWA headquarters lacks information on what funding was made available and remains unobligated to states for specific events.

Because Emergency Relief Program funding is not subject to the annual limits that the regular federal-aid highway program is, states have an incentive to retain as much emergency relief funding as possible by not returning unused funds. The lack of information on the amount of funds that could be made available for specific events could prevent FHWA from verifying whether allocations provided to states are still needed or may be withdrawn and used to meet current needs.

Table 2. Emergency Relief Program Obligations and Expenditures, Fiscal Years 2001 through 2006

Dollars in millions							
			Fiscal year				
	2001	2002	2003	2004	2005	2006	Total
Obligations	$360	$67	$195	$1,082	$1,991	$758	$4,454
Expenditures	314	65	191	1,072	1,803	667	4,113
Unexpended balance	46	2	5	10	188	91	341
Percent expended	87%	97%	98%	99%	91%	88%	92%

Source: GAO analysis of FHWA financial data.

Note: The amounts provided for each fiscal year were rounded to the nearest million and may not add to the total amount provided due to rounding.

In Addition to Unused Allocations, Obligated Funds Remain Unexpended

In addition to the unused allocations, substantial amounts of obligated emergency relief funding have not been expended. About $642 million in emergency relief funding obligated for states from fiscal years 2001 through 2010 remains unexpended as of May 2011—including about $341 million in emergency relief funds obligated from fiscal years 2001 through 2006.[50] In total for the Emergency Relief Program, 8 percent of all funding obligated from fiscal years 2001 through 2006 has yet to be expended (see table 2).

Almost half of the unexpended balance from fiscal years 2001 through 2006 is for projects in response to several extraordinary events that occurred during those years, including the September 11, 2001, terrorist attacks in New York and Gulf Coast Hurricanes Katrina, Rita, and Wilma in 2005. Specifically, about $45 million of the $46 million that remains unexpended for fiscal year 2001 is for repair projects to facilities around the World Trade Center site in New York City. Of the $188 million that remains unexpended for fiscal year 2005, about $118 million is for projects in Louisiana in response to Hurricane Katrina. As of the end of May 2011, FHWA obligated about $952 million to 155 emergency relief projects in Louisiana for this event and has since made reimbursements to the state for all but 1 of these projects, providing approximately 88 percent of the amount obligated.

Although substantial unexpended obligated funding remains, FHWA lacks information to determine the amount that is unneeded and could be deobligated because there is no time frame for closing out completed emergency relief projects. FHWA division officials in New York and Texas reported that many emergency relief projects are administered by local public agencies, including towns and counties, and these entities are often slow to process their reimbursement requests through the state department of transportation. As such, FHWA lacks information on the status of these projects and whether projects are ongoing or have been completed. For example, in Texas, 28 of 30 projects since 2007 included in our file review were listed as active in FHWA's national database, FMIS. However, according to Texas Department of Transportation (TXDOT) officials, construction on 23 of the 28 active projects was in fact completed and waiting to be closed out. FHWA division office officials reported that FMIS is not a project management system and does not provide the actual status of the construction of projects. As such, states may have completed some emergency relief projects but not processed reimbursement requests from local public agencies or completed final project financial audits. Projects remain active in FMIS until final vouchers have been processed to reimburse states. DOT's Office of Inspector General and external independent auditors have both identified inactive or unexpended obligations as a significant concern within FHWA.[51] Without clear time frames for states to close out completed emergency relief projects, FHWA lacks important information on the status of projects and whether unexpended project funds are no longer needed and may be deobligated to be made available for other emergency relief projects.

Prior Concerns about Project Eligibility Have Yet to Be Addressed

FHWA has yet to address our longstanding concern about, and our 2007 recommendation for addressing, the use of emergency relief funds to finance projects that have grown in scope beyond the original intent of the program, which is to restore damaged facilities to predisaster conditions.[52] In 1996, we questioned FHWA's decision to use more than $1 billion in emergency relief funds to replace the Cypress Viaduct in Oakland, California, which collapsed as a result of the Loma Prieta earthquake in October 1989.[53] FHWA engineers initially estimated that replacing the destroyed structure along its predisaster alignment would cost $306 million. In response to public concern, Caltrans identified several alternative alignments that it studied in a 2-year environmental review. In 1991, Caltrans and FHWA decided to replace the destroyed 1.5-mile structure, which had bisected a residential area, with a new 5- mile structure running through active rail yards. This cost estimate later increased to more than $1.1 billion at the time of our 1996 report—an increase of almost $800 million from FHWA's initial estimate of $306 million to restore the facility to its predisaster condition. As such, we questioned whether the improvements and costs resulting from the significant relocation and changes in scope should have been funded through the Emergency Relief Program rather than the regular federal-aid highway program. We recommended that FHWA modify its guidance to clearly define what costs can be funded through the Emergency Relief Program, particularly when an environmental review recommends improvements or changes to the features of a facility from its predisaster condition in a manner that adds costs and risks to the project.

Although FHWA took steps to address our 1996 recommendation,[54] in 2007, we reported several cases in which the Emergency Relief Program was used to fund large projects—such as a bridge replacement or road relocation—that were either delayed, had grown in project scope and costs, or went beyond the original intent of the Emergency Relief Program to restore damaged facilities to predisaster conditions.[55] First, we noted that relocating California S.R.1 at Devil's Slide could have been addressed through the state's regular federal-aid highway program, rather than through the Emergency Relief Program. If the regular federal-aid highway program had been used, the project would not have been eligible for 100 percent federal funding,[56] and the federal government would have saved an estimated $73 million.[57] Second, we reported that the reconstruction of the U.S. Highway 90 Biloxi Bay Bridge in Mississippi— which was destroyed in August 2005 during Hurricane Katrina— grew in scope and cost by $64 million as a result of community concerns. Specifically, in response to a concern raised by a local shipbuilder about the proposed height of the new bridge, Mississippi department of transportation expanded the scope of the bridge reconstruction to increase the bridge height to allow for future ships to pass under the bridge. The original design was to provide an 85-foot clearance at a cost of $275 million, but this scope was expanded to its current design to provide a 95-foot clearance at a cost of $339 million.[58]

FHWA has clarified its definition of an eligible damage site as we recommended in 2007, through its revisions to its *Emergency Relief Manual* in 2009. Specifically, FHWA's 2009 revisions clarified that grouping damages to form an eligible site based solely on a political subdivision (i.e., county or city boundaries) should not be accepted. This change addressed our concern that FHWA division offices had different interpretations of what constituted a site, such that damage sites that were treated as eligible for emergency relief in one state may have not been eligible in another state.

INCOMPLETE INFORMATION IN EMERGENCY RELIEF PROJECT FILES IN THREE STATES RAISES CONCERNS ABOUT FHWA'S ELIGIBILITY DECISIONS AND PROGRAM OVERSIGHT

Documentation for Many Project Files We Reviewed Was Missing, Incomplete or Inconsistent

In our review of 83 selected emergency relief project files in three FHWA division offices, we found that many of the project files reviewed did not contain documentation called for in the *Emergency Relief Manual* to support FHWA decisions that projects met program eligibility requirements.[59] Of the 83 projects in our review (totaling about $198.5 million in federal funds), 81 projects (about $192.8 million in federal funds) had at least one instance of missing or incomplete documentation. As a result of this missing information, we were unable to determine the basis of FHWA's eligibility decisions for many of the projects in our file review.

The *Emergency Relief Manual* directs FHWA division offices to maintain files containing information on the methods used to evaluate disasters and FHWA's assessment of damages and estimates of cost. According to the Emergency Relief Program regulations, program data should be sufficient to identify the approved disaster and permit FHWA to determine the eligibility of the proposed work. 60 In our file review, we identified several areas of concern with FHWA's eligibility determinations based on missing, incomplete, or inconsistent documentation, as illustrated in table 3 and described below (see app. III for detailed results of our file review).

Missing or Incomplete Detailed Damage Inspection Reports

Forty-seven of 83 project files (57 percent) lacked documentation for on-site damage inspections. In particular, they did not include a detailed damage inspection report (DDIR) or the DDIR was not complete. According to the *Emergency Relief Manual*, on-site detailed damage inspections are conducted by the applicant or a state department of transportation representative if the applicant is a local public agency, and an FHWA representative, if available, to determine the extent of damage, scope of repair work, preliminary estimate of the repair cost, and whether a project is eligible for emergency relief funding.

Table 3. Instances of Missing or Incomplete Project Documentation

Areas of concern regarding eligibility	New York	Texas	Washingtonstate	Total instances
Missing or incomplete detailed damage inspection reports	16 of 22	28 of 28	3 of 33	47 of 83
Missing repair cost estimates	12 of 22	14 of 28	16 of 33	42 of 83
Missing or incomplete emergency repair completion dates to support 100 percent	18 of 18[a]	0 of 17	21 of 23	39 of 58[b]
Missing betterment justification	1 of 1	0 of 4	5 of 10	6 of 15c

Source: GAO analysis of FHWA Emergency Relief Program documentation.

[a] The 18 projects with emergency repairs in New York provided only a month and year for the completion of repairs. Without a specific date we could not confirm completion was within 180 days of the disaster occurrence. These 18 projects are indicated as partially complete for having a completion date on file for 100 percent funded work in app. III, figure 6.

[b] Fifty-eight of the 83 projects in our review included emergency repairs and were approved to receive 100 percent federal share funding for either all or a portion of the total project cost.

[c] Fifteen of the 83 projects in our review were identified as betterments in project documentation or by FHWA division office officials with whom we spoke.

FHWA provides its division offices with a DDIR form that states may use to document their inspections and provide critical information necessary for determining project eligibility, such as a listing of preliminary repair cost estimates for equipment, labor, and materials for both emergency and permanent repairs. Without such information on file for some projects, we could not confirm that FHWA had that information to make emergency relief project eligibility determinations.

These documents may be missing due to lack of clear requirements from FHWA. FHWA requires documented on-site damage inspections but does not have a clear requirement for how states submit the inspections to FHWA officials or for how they approve inspection reports; as a result, the three division offices we visited applied the *Emergency Relief Manual* guidelines differently. For example, none of the 28 project files we reviewed in Texas included a DDIR because FHWA's Texas division office relies instead on a "program of projects," which is a spreadsheet of all projects requesting emergency relief funds. In response to a draft version of this report, FHWA's Office of Program Administration explained that state departments of transportation may use any format to submit the data necessary for FHWA to make an eligibility determination. FHWA's Texas division officials stated that they find the program of projects useful and believed it to be an FHWA requirement; however, we found that the *Emergency Relief Manua*l guidance was ambiguous and did not directly state that this document can be used in place of DDIRs. One section the *Emergency Relief Manual* indicates that the state department of transportation is to submit the program of projects to the FHWA division office, but it also states that the program of projects should relate the damage to that described in the DDIRs. Furthermore, the manual suggests in an appendix that the program of projects is actually a package of all DDIRs resulting from the detailed damage inspections.

In addition, our file review found that the project descriptions in the program of projects did not always provide the detailed information regarding damages and proposed repairs outlined in the *Emergency Relief Manual* and found on a DDIR. For example, for one Texas project totaling close to $1.7 million in both emergency and permanent repairs, the project description was the same for both emergency and permanent repairs and did not indicate what specific repair activities were conducted for each repair type. Differentiation between emergency and permanent repairs is important because emergency repairs are eligible for a higher federal share and do not require prior FHWA authorization. Without documentation showing a clear distinction between the emergency and permanent repairs—information that should be identified and documented on a DDIR per program guidance—we could not determine the basis for FHWA's decision that this project met the eligibility requirements for both repair types. Overall, we found the program of projects was less useful than the DDIR for evaluating the full range of information necessary to determine the basis for FHWA's eligibility determinations.

Missing Repair Cost Estimates

We found that about half of the projects in our sample (42 of 83) did not include repair cost estimates. The *Emergency Relief Manual* states that at a minimum the division office's project file should contain copies of the FHWA field engineer's assessments on damage and estimates of cost.

Officials in each of the FHWA division offices that we visited reported that the state's department of transportation is responsible for preparing repair cost estimates, but that FHWA area engineers also conduct some on-site inspections to verify the cost estimates provided. In total, 42 projects in our sample did not include any repair cost estimates; thus, we could not confirm that FHWA officials had this information to make eligibility determinations for those projects. For example, a portion of two projects in our sample for emergency and permanent repairs was to remove sand from drainage ditches and was initially approved by the FHWA Texas division office for reimbursement of up to $1.3 million, although the project file included no repair cost estimate for any of the work associated with the project. Additionally, no information was available in the project file to explain the FHWA Texas division office's decision to later approve a nearly 40 percent increase from $1.3 million to the final approved amount of $1.85 million. In responding to a draft of this report, DOT stated that the cost of the project increased because more sand was removed from the drainage ditches than originally estimated. However, no documentation of this change was included in FHWA's project files.

FHWA officials reported that the division office in Texas reviews a sample of preliminary cost estimates based on risk, among other factors, prior to making any eligibility decisions. According to the officials, FHWA's Texas division office reviewed preliminary cost estimates of at least 10 of the 30 projects included in our file review before determining eligibility. The officials also reported that this sampling approach is consistent with FHWA's stewardship agreement with TXDOT and the fact that states have assumed oversight responsibility for design and construction of many federal-aid highway projects, including emergency relief projects.[61] FHWA also reported that TXDOT's oversight responsibilities do not extend to determining whether particular projects are eligible for federal funds. Furthermore, the *Emergency Relief Manual* states that Emergency Relief Program eligibility determinations reside with FHWA, and estimated repair costs should be documented to determine eligibility. As such, the practice of reviewing a sample of preliminary cost estimates does not appear to be consistent with the requirements in the *Emergency Relief Manual*, and as a result, we could not determine the basis of FHWA's eligibility decisions for those project cost estimates it did not review.

We found other cases in which cost increases were not documented according to the internal policies established by each of the division offices we visited. In New York and Texas, FHWA division officials stated they require additional documentation to justify cost increases of 25 percent or more. In Washington state, FHWA division office officials stated they require additional documentation if costs increase by 10 percent or more. Yet 14 percent of the project files we reviewed (12 of 83) showed total cost increases that exceeded the limits established by the three division offices and no additional documentation was on file to support the increases.[62]

Missing or Incomplete Emergency Repair Completion Dates to Support Eligibility for 100 Percent Federal Funding

The majority of the emergency repair project files that we reviewed did not include documentation demonstrating that emergency repairs were completed within 180 days from the event to be eligible for 100 percent federal reimbursement.[63] Fifty-eight of the 83 projects we reviewed included emergency repairs approved to receive 100 percent federal funding reimbursement if repairs were completed within 180 days of the event occurrence. However, 39 of the 58 (67 percent) did not have documentation on file to show the completion date of those repairs (see table 3). In total, only 14 of 58 (24 percent) emergency repair projects provided a completion date that was within 180 days of the event's occurrence. For the majority (39 of 58) of projects, we were unable to confirm whether the emergency repairs were completed within 180 days and whether these projects were eligible to receive 100 percent federal reimbursement.

The *Emergency Relief Manual* does not clearly establish specific requirements that states demonstrate and FHWA verify the date of project completion. As such, FHWA lacks a standardized process for verifying the completion of emergency repairs within 180 days on projects for which it does not exercise full oversight. By law, states assume oversight responsibility for the design and construction of many federal-aid highway projects,[64] including the vast majority of emergency relief projects in the three divisions we visited. As such, the states—rather than FHWA—were responsible for conducting final inspections of emergency relief projects. States are required to conduct a final inspection for all federal-aid highway projects under state oversight, and these inspections could be useful to determine federal share eligibility of emergency repairs if they provide project completion dates. While officials in each of the three state departments of transportation told us that they conduct final inspections of emergency repairs, we found only two final inspection reports prepared by states in FHWA's records to confirm the completion of emergency repairs within the required time frame. In addition, when we reviewed final inspection reports from one of the state departments of transportation in our review, we were frequently unable to verify completion dates. Specifically, 11 of the 12 final inspections performed by officials at New York State Department of Transportation for projects in our review did not include project completion dates. Although the *Emergency Relief Manual* states that FHWA division offices reserve the right to conduct a final inspection of any emergency relief project, only the FHWA Texas division reported conducting spot inspections for a sample of emergency relief projects.[65]

In commenting on a draft of this report, DOT stated that the FHWA New York state division office uses other means to verify completion of emergency repairs within 180 days. According to DOT, the state often submits its DDIRs to FHWA after emergency repairs are completed, which allows FHWA to verify the eligibility and completion of an emergency repair when it reviews the DDIR. DOT reported that the FHWA division office does not sign the DDIR until it confirms the work is completed, and that its signature indicates verification that the work was performed within the required time frame. However, our file review found that 14 of the 18 emergency repair projects in New York that were approved for 100 percent federal funding did not have an FHWA signature on the DDIR.

In addition to a lack of documentation, we found eight instances in which permanent repair projects may have incorrectly received 100 percent federal share reimbursement.[66] According to the *Emergency Relief Manual*, absent specific legislative approval, permanent repair work is not to be considered emergency repair work even if it is completed within 180 days. However, we found instances in which projects were determined to be permanent repairs based on information in the project files, but were later authorized to receive 100 percent federal share. For example, in one project in our review, FHWA's Washington state division office approved permanent repairs to a state highway for $2.6 million in estimated damages caused by a landslide. Our review of FHWA financial records for this project indicates that FHWA later authorized a federal reimbursement of $5.3 million, roughly 99 percent of the total project cost of nearly $5.4 million. FHWA Washington state division officials reported that this project was considered to be a permanent repair performed as an incidental part of emergency repair work.[67] However, the project files did not include any emergency repair work to accompany the approved permanent repairs. According to these officials, the FHWA Washington state division interpreted the 2003 version of the *Emergency Relief Manual* as allowing incidental permanent work to be funded at 100 percent federal share either with or as emergency repair work. However, the manual states that during the 180 day period following the disaster, permanent repair work is reimbursed at the normal pro rata share unless performed as an incidental part of emergency repair work. As such, based on the program guidance, this project should have been reimbursed at 86.5 percent federal share.

Missing and Inconsistent Support for Betterments

A primary purpose of the Emergency Relief Program is to restore highway facilities to predisaster conditions, not to provide improvements or added protective features to highway facilities. However, according to FHWA regulations and the *Emergency Relief Manual*, such improvements may be considered eligible betterments if the state provides economic justification, such as a benefit-cost analysis that weighs the cost of the betterment against the risk of eligible recurring damage and the cost of future repair through the Emergency Relief Program.[68] In our file review we identified two areas of concern regarding betterments, including instances of missing documentation of benefit-cost analyses:

- *Lack of documentation of required benefit-cost analyses*. Six of the 15 projects (40 percent) identified as betterments in our review did not contain the required benefit-cost analyses in their files to justify the betterment. 69 As a result we were unable to determine the basis on which FHWA approved these six betterments. We also found one instance in which the benefit-cost analyses used to justify an approved betterment did not meet Emergency Relief Program requirements. Specifically, FHWA's New York division office approved a betterment of almost $1.6 million to repair and improve a damaged roadway and shoulder caused by an April 2007 storm. However, we found that the report prepared to justify the betterment did not weigh the cost of the proposed betterment against the risk of future damages and repair costs to the Emergency Relief Program, as required by program regulations. Consequently, we were unable to determine the basis on which FHWA approved the $1.6 million betterment.

- *Lack of documentation indicating whether projects include betterments.* We found that it was often difficult to determine which projects included betterments, as FHWA lacks a standard process for where and how betterments should be identified in project documentation. The *Emergency Relief Manual* states that betterments must receive prior FHWA approval and that further development of contemplated betterments should be accomplished with FHWA involvement, necessitating that proposed betterments are specifically identified. We found eight project files with indications that the projects may have included betterments that were not identified explicitly in project documentation or by FHWA officials.[70] For example, following the completion of emergency repairs to remove debris and protect a bridge against erosion caused by a landslide, the FHWA Washington state division office approved an additional $3.7 million in permanent repairs in response to continued erosion and movement of the hillside. The documentation in the project file indicated that this permanent work was added to stabilize the slide area in anticipation of future flooding. According to officials from the FHWA Washington state division, this slide stabilization project was a betterment, but the project file did not contain documentation to indicate that this project was in fact a betterment.

FHWA provides considerable discretion to its division offices to tailor the Emergency Relief Program within states and lacks a standard mechanism to specifically identify whether a project includes a betterment. FHWA's Office of Asset Management has developed an Economic Analysis Primer for FHWA division offices to use when evaluating benefit-cost analyses for federal-aid program projects. However, neither the *Emergency Relief Manual* nor the Economic Analysis Primer provide sample benefit-cost analyses or specific guidance on what information should be included in the benefit-cost analysis to demonstrate that the proposed betterment will result in a savings in future recurring repair costs under the Emergency Relief Program. Because we had found betterments without documentation of the required benefit-cost analyses on file and identified possible betterments that were not explicitly identified as such, we could not confirm that federal funds were being reimbursed in accordance with the requirements of the Emergency Relief Program. Further, absent specific guidance for identifying and approving betterments to its division offices, FHWA cannot be assured that the Emergency Relief Program is being administered consistently.

CONCLUSION

The federal government plays a critical role in providing financial assistance to states in response to natural disasters and other catastrophic events. Given the costs of these events and the significant fiscal challenges facing both states and the federal government, it is increasingly necessary that federal financial support be delivered in an effective, transparent, and accountable manner so that limited funds are put to their best use. FHWA's stewardship of the Emergency Relief Program could be better structured to meet that necessity.

First, because some emergency relief projects can be delayed for many years due to environmental or community concerns and projects can grow significantly in scope and cost, the federal government faces the risk of incurring long-term costs for such projects. FHWA has limited tools to control its exposure to the costs of older events and ensure that as projects grow in scope and cost that they do not go beyond the original intent of the program, which is to assist states to restore damaged facilities to their predisaster conditions. Once an event has been approved for emergency relief by FHWA, the Emergency Relief Program as currently structured does not limit the time during which states may request additional funds and add projects, which increase the size of FHWA's backlog list. Because Emergency Relief Program funding is not subject to the annual limits of the regular federal-aid highway program, states have an incentive to seek as much emergency relief funding as possible. Consequently, without reasonable time limits for states to submit funding requests for such older events, FHWA's ability to anticipate and manage future costs to the Emergency Relief Program is hindered, as is Congress' ability to oversee the program. Furthermore, without specific action by FHWA to address the recommendation from our 2007 report that it revise its emergency relief regulations to tighten eligibility criteria, the Emergency Relief Program will continue to face the risk of funding projects with scopes that have expanded beyond the goal of emergency relief and may be more appropriately funded through the regular federal-aid highway program.

Second, while FHWA has taken some important steps in response to our 2007 report to manage program funding by withdrawing unobligated balances from states, it faces challenges in tracking allocations that have been provided to states. In particular, because FHWA division offices have allowed states to transfer unobligated allocations from an existing event to new events, and because FHWA headquarters is not tracking which divisions have done so, FHWA headquarters does not have the information needed to identify and withdraw all unneeded funds. In addition, without time frames to expedite the close-out of completed emergency relief projects, FHWA lacks useful information to help determine whether obligated but unexpended program funds are no longer needed and could be deobligated.

Finally, the fact that we could not determine the basis of FHWA's eligibility decisions in three states on projects costing more than $190 million raises questions about whether emergency relief funds are being put to their intended use and whether these issues could be indicative of larger problems nationwide. While federal law allows states to assume oversight over design and construction of much of the federal-aid highway program, including many emergency relief projects, FHWA is ultimately responsible for ensuring that federal funds are efficiently and effectively managed and that projects receiving scarce emergency relief funds are in fact eligible. This is especially important in light of the fact that emergency relief funds have been derived principally from general revenues in recent years and that the funds that states receive are above and beyond the funding limits for their regular federal-aid highway program funds. Without clear and standardized procedures for divisions to make and document eligibility decisions—including documenting damage inspections and cost estimates, verifying and documenting the completion of emergency repair projects within the required time frame, and evaluating information provided to justify proposed betterments—FHWA lacks assurance that only eligible projects are approved, and that its eligibility decisions are being made and documented in a clear, consistent, and transparent manner.

Recommendations for Executive Action

To improve the accountability of federal funds, ensure that FHWA's eligibility decisions are applied consistently, and enhance oversight of the Emergency Relief Program, we recommend that the Secretary of Transportation direct the FHWA Administrator to take the following four actions:

- Establish specific time frames to limit states' ability to request emergency relief funds years after an event's occurrence, so that FHWA can better manage the financial risk of reimbursing states for projects that have grown in scope and cost.
- Instruct FHWA division offices to no longer permit states to transfer unobligated allocations from a prior emergency relief event to a new event so that allocations that are no longer needed may be identified and withdrawn by FHWA.
- Establish clear time frames for states to close out completed projects in order to improve FHWA's ability to assess whether unexpended program funds are no longer needed and could be deobligated.
- Establish standardized procedures for FHWA division offices to follow in reviewing emergency relief documentation and making eligibility decisions. Such standardized procedures should include:
- clear requirements that FHWA approve and retain detailed damage inspection reports for each project and include detailed repair cost estimates;
- a requirement that division offices verify and document the completion of emergency repairs within 180 days of an event to ensure that only emergency work completed within that time frame receives 100 percent federal funding; and
- consistent standards for approving betterments, including guidance on what information the benefit-cost analyses should include to demonstrate that the proposed betterment will result in a savings to the Emergency Relief Program, and a requirement that FHWA approval of funding for betterments be clearly documented.

Agency Comments and Our Evaluation

We provided a draft of this report to DOT for review and comment. DOT officials provided technical comments by email which we incorporated into the report, as appropriate. In response to our finding that the Emergency Relief Program lacks a time limit for states to submit emergency relief funding requests, and our recommendation to establish specific time frames to limit states' ability to request emergency relief funds years after an event's occurrence, DOT noted that the program does include general time frames for states to submit an application and have work approved. We incorporated this information into the final report; however, since a state's list of projects may be amended at any time to add new work, we continue to believe that FHWA's ability to anticipate and manage future costs to the Emergency Relief Program is hindered absent specific time frames to limit states' requests for additional funds years after an event's occurrence. Such time frames would provide FHWA with an important tool to better manage program costs.

DOT also commented that its ability to control the costs of some of the projects cited in the report that have grown in scope and cost over the years is limited in some cases by the fact that DOT received statutory direction from Congress to fund these projects. For example, Congress directed FHWA to provide100 percent federal funding for all emergency relief projects resulting from Hurricane Katrina in 2005. We incorporated additional information to recognize this statutory direction; however, a determination by Congress that a particular event should qualify for relief under the Emergency Relief Program, or for other individual actions, does not relieve FHWA of its stewardship and oversight responsibilities. Except as Congress otherwise provides, this includes its responsibility to determine whether enhancements to projects or betterments are consistent with its regulations and the intent of the Emergency Relief Program to restore damaged facilities to predisaster conditions. We continue to believe that, as a steward of public funds, FHWA generally has the discretion to take reasonable steps to limit the federal government's exposure to escalating costs from projects that grow in scope over time.

Phillip R. Herr
Director, Physical Infrastructure Issues

APPENDIX I: OBJECTIVES, SCOPE AND METHODOLOGY

To identify Emergency Relief Program funding trends since our 2007 report,[1] we reviewed federal statutes, including supplemental appropriations to the Emergency Relief Program made since 2007, and Federal Highway Administration (FHWA) documentation on annual funding authorizations to the program. We also reviewed FHWA data on emergency relief funds allocated to states in response to emergency relief events from fiscal years 2007 through 2010, as provided by FHWA's Office of Program Administration. We interviewed FHWA officials in the Office of Program Administration to gather specific information on how data on allocations was collected and we also reviewed FHWA financial data on total allocations to states from FHWA's fiscal management information system (FMIS). We interviewed officials from FHWA Federal Lands Highway, FHWA's North Dakota Division Office, and the North Dakota state department of transportation concerning funding and project activities for the Devils Lake, North Dakota, emergency relief projects. To gather additional information on the Devil's Slide project in California, we interviewed the FHWA California division office and reviewed information on the estimated project costs.

To identify key changes to the Emergency Relief Program implemented in response to concerns raised in our 2007 report, we reviewed recommendations made to FHWA in our 2007 report and FHWA Emergency Relief Program regulations and guidance, including FHWA's *Emergency Relief Manual*, as revised in 2009. We compared information in the current version of the *Emergency Relief Manual* with information in the previous version to determine which elements were revised. We interviewed FHWA officials in the Office of Program Administration to determine why specific changes were made, and we interviewed officials in three FHWA division offices to determine how program changes were implemented. To corroborate information provided by FHWA regarding its process of withdrawing unused Emergency Relief Program funds from states, we reviewed FMIS data on the emergency relief funds that were allocated among all states and territories, obligated to specific projects, and the remaining unobligated balance for all active Emergency Relief Program codes as of May 31, 2011. To determine other amounts of program funding that remained unused, we reviewed data in FMIS on the amount of emergency relief funding obligated to specific projects and expended by all states and territories for events occurring from fiscal years 2001 through 2010. We provided FHWA officials with our methodology for gathering data from FMIS to ensure that our data queries were accurate. To ensure the reliability of data collected in FMIS we interviewed FHWA officials on the procedures used by FHWA and states' departments of transportation to enter and verify financial information entered into FMIS. We found these data to be sufficiently reliable for our purposes.

To determine the extent to which selected emergency relief projects were awarded in compliance with program eligibility requirements, we reviewed federal statutes and regulations, and FHWA guidance on emergency relief eligibility requirements. We selected a nongeneralizable sample of state department of transportation and FHWA division offices in three states—New York, Texas, and Washington state. The states selected are not representative of the conditions in all states, the state departments of transportation, or FHWA division offices, but are intended to be examples of the range of practices and projects being funded by the Emergency Relief Program across the country. These states were selected based on several criteria:

1) The overall amount of emergency relief funding allocated to a state from fiscal years 2007 through 2010, to identify those states that were allocated the most funding (at least $15 million) over that period, based on allocation data provided by FHWA headquarters.
2) Frequency of funding requests to identify those states that requested funds for three or more fiscal years from 2007 through 2010.
3) The occurrence of an eligible emergency relief event since FHWA updated its *Emergency Relief Manual* in November 2009. For our purposes, we used emergency relief eligible events beginning October 1, 2009, as a proxy for identifying states with emergency relief events since the November 2009 manual update.

A total of 10 states met all three criteria. We narrowed our selection down by eliminating those states that experienced outlier events, such as North Dakota's reoccurring basin flooding at Devils Lake and the catastrophic failure of the Interstate 35 West bridge in Minnesota. We judgmentally selected New York, Texas, and Washington state to reflect a geographic dispersion of states.

We reviewed a sample of emergency relief project files in the FHWA division office in each of these states to determine whether the project files included required or recommended documentation cited in federal statute, regulations, and FHWA program guidance. Such documentation included the President or state governors' proclamation of a disaster, detailed damage inspection reports, cost estimates for repairs, photographs of the damage, and other information. Across the three division offices, we selected a nongeneralizable sample of 88 Emergency Relief Program files out of a total universe of 618 project files for emergency relief projects approved by FHWA in those states from fiscal years 2007 through 2010. Among the 88 projects in our review, 5 projects had been withdrawn by states as FHWA had determined them ineligible for emergency relief funds, or they were reimbursed through a third party insurance settlement, bringing the total number of projects reviewed to 83. The project files we reviewed represented approximately 67 percent of all emergency relief funds obligated to those states during that time period.[2] Those projects were selected based on the following criteria:

1) All projects with more than $1 million in obligated federal funds between fiscal years 2007 and 2010, including a mix of active and closed projects and various event or disaster types.
2) Projects with more than $1 million in obligated federal funds for events from fiscal years 2001 through 2006 on the list of formal emergency relief funding requests as of March 7, 2011, that were either currently active or were completed more than five years after the event occurred.
3) Projects that had other characteristics that we determined to warrant further review, such as events with $0 amounts listed in FHWA's FMIS database for total cost or which had expended relatively small amounts of funding compared with the obligated amounts in FMIS.

Prior to our site visits, we requested that the division offices provide all documentation they maintain for each of the projects selected in our sample. We reviewed all the documentation provided during our site visits, and requested follow-up information as necessary. In conducting our file review, a GAO analyst independently reviewed each file and completed a data collection instrument to document the eligibility documentation that was included for each file. A second reviewer independently reviewed the file to verify whether the specific information identified by the first reviewer was present in the file. The analysts met to discuss and resolve any areas of disagreement until a consensus was reached on whether the required information was included in the file. To gather additional information on the project files we reviewed and the procedures used to manage and oversee emergency relief projects, we interviewed officials in the FHWA division offices and the departments of transportation in our three selected states. We provided the results of our file review to FHWA for their comment and incorporated their responses as necessary within our analysis.

Lastly, we contacted state and local audit organizations through the National Association of State Auditors, Comptrollers, and Treasurers for the three states we reviewed, as well as North Dakota, to obtain reports or analyses that were conducted on FHWA's Emergency Relief Program. None of the states in our review had conducted substantive work on the Emergency Relief Program.

We conducted this performance audit from November 2010 to November 2011 in accordance with generally accepted government auditing standards. Those standards require that we plan and perform the audit to obtain sufficient, appropriate evidence to provide a reasonable basis for our findings and conclusions based on our audit objectives. We believe that the evidence obtained provides a reasonable basis for our findings and conclusions based on our audit objectives.

APPENDIX II: SUMMARY OF EMERGENCY RELIEF FUNDING FOR PROJECTS AT DEVILS LAKE, NORTH DAKOTA

Devils Lake in North Dakota lies in a large natural basin and lacks a natural outlet for rising water to flow out of the lake. Starting in the early 1990s, the lake level has risen dramatically—nearly 30 feet since 1992— which has threatened the roadways near the lake which were built in the 1930s and 1940s when lake water levels were lower. In April 2000, FHWA issued a memorandum that authorized raising the roads at Devils Lake in response to a predicted rise in the water level of the lake that was within 3 feet of causing inundation, as forecasted by the National Weather Service or U.S. Geological Survey. This allowance to repair roadways prior to damages incurred by an event is a unique provision for the FHWA Emergency Relief Program, which otherwise funds only post-disaster repair or restoration. The basin flooding events at Devils Lake also precipitated a related problem at Devils Lake, as some communities around the lake plugged culverts under roadways to impound rising water and protect property from flooding, which increased the roadways' risk of failure. These roads were subsequently referred to as "roads-acting-asdams" which required additional improvements to ensure their structural integrity to serve as dams.

Devils Lake projects involve multiple stakeholders, depending on the location and type of roadway. FHWA's North Dakota division office is responsible for overseeing the Emergency Relief Program projects administered by North Dakota department of transportation. FHWA's Office of Federal Lands Highway is responsible for the oversight of the Emergency Relief on Federally Owned Roads program, which covers projects on the Spirit Lake Tribe Indian Reservation.

The Central Division of Federal Lands Highway leads the overall coordination among the federal, state, and local agencies. FHWA reported that the two FHWA offices are working together to address the roads-acting-as dams projects which affect state highways and roads on the Sprit Lake Tribe Indian Reservation. The North Dakota department of transportation and the Spirit Lake Tribe are responsible for administering the construction projects on their respective roads.

To ensure the integrity of the roads at Devils Lake, Congress included funding provisions in Safe, Accountable, Flexible, Efficient Transportation Equity Act: A Legacy for Users (SAFETEA-LU) to raise the roadways and make improvements to roads-acting-as-dams. Through SAFETEA-LU, Congress authorized up to $10 million of Emergency Relief Program funds to be expended annually, up to a total of $70 million, for work in the Devils Lake region of North Dakota to address the roads-acting-as-dams situation.[1] These funds are known as section 1937 funds for the provision in SAFETEA-LU which authorized them. In the absence of other authority, this $10 million must come out of the $100 million annual authorization of contract authority that funds the Emergency Relief Program, effectively reducing the annual emergency relief funding available to other states to $90 million. SAFETEA-LU also included language that exempted the work in the Devils Lake area from the need for further emergency declarations to qualify for emergency relief funding.[2]

According to a June 24, 2011, FHWA policy memo, the final allocation of section 1937 funds was made on March 16, 2011, and the $70 million limit has been reached. Although rising water levels at Devils Lake are expected to continue into the future, no further federal-aid highway funds are eligible to raise roads-acting-as-dams or to construct flood control and prevention facilities to protect adjacent roads and lands.

APPENDIX III: RESULTS OF GAO'S FILE REVIEW OF EMERGENCY RELIEF PROJECT DOCUMENTATION AVAILABLE IN THREE FHWA DIVISION OFFICES

Figure 6 represents the results of our review of 88 selected project files from FHWA's division offices in New York, Texas, and Washington state. Our data collection instrument was used to collect the values for each field during our file review, and that information was summarized and analyzed by at least two GAO analysts (see app. I for a complete discussion of our file review methodology).

New York

GAO project number	DDIR on file	FHWA engineer signature on DDIR	State/ local engineer signature on DDIR	Repair cost estimate	Repair type	Completion date recorded in file for 100 percent funded work	Dates provided show 100% funded work completed in 180 Days	FHWA verification of completion within 180 days	Betterment	Benefit-cost analysis on file for betterment
NY-1	Y	N	Y	N	E/P	Y	DK[b]	N	N	
NY-2	Y	N	Y	Y	E/P	Y	DK[b]	N	Y	N
NY-3	Y	N	N	N	DK	Y	DK[b]	N	N	
NY-4	Y	N	Y	N	E/P	Y	DK[b]	N	N	
NY-5	Y	N	Y	Y	E	Y	DK[b]	N	N	
NY-6	Y	Y	Y	Y	P				N	
NY-7	Y	N	Y	Y	E/P	Y	DK[b]	N	N	
NY-8	Y	N	Y	Y	E	Y	DK[b]	N	N	
NY-9	Y	N	Y	Y	DK				N	
NY-10	Y	N	N	N	E/P	Y	DK[b]	N	N	
NY-11	Y	N	Y	Y	E/P	Y	DK[b]	N	DK[a]	
NY-12	Y	N	N	N	E/P	Y	DK[b]	N	N	
NY-13	Y	N	N	N	E	Y	DK[b]	N	N	

- Y — Documentation was on file (Y=yes)
- Y — Documentation was on file but only partially complete
- DK — Unable to determine the value based on documentation on file (DK=don't know)
- N — Documentation was not on file (N=no)
- Not applicable to that project
- E — Emergency repair
- P — Permanent repair
- I — Incidental permanent repair

GAO project number	DDIR on file	FHWA engineer signature on DDIR	State/ local engineer signature on DDIR	Repair cost estimate	Repair type	Completion date recorded in file for 100 percent funded work	Dates provided show 100% funded work completed in 180 Days	FHWA verification of completion within 180 days	Betterment	Benefit-cost analysis on file for betterment
NY-14	Y	N	Y	Y	E	Y	DK[b]	N	N	
NY-15	Y	N	N	N	E	Y	DK[b]	N	N	
NY-16	Y	N	Y	Y	E	Y	DK[b]	N	N	
NY-17	Y	N	Y	N	E	Y	DK[b]	N	N	
NY-18	Y	Y	Y	N	E	Y	DK[b]	N	N	
NY-19	Y	Y	Y	N	E	Y	DK[b]	N	N	
NY-20	Y	Y	Y	Y	E/P	Y	DK[b]	N	N	
NY-21	Y	Y	Y	N	E	Y	DK[b]	N	N	
NY-22	Y	Y	Y	N	DK	Y	DK[b]	N	N	

Y Documentation was on file (Y=yes)

Y Documentation was on file but only partially complete

DK Unable to determine the value based on documentation on file (DK=don't know)

N Documentation was not on file (N=no)

Not applicable to that project

E Emergency repair

P Permanent repair

I Incidental permanent repair

Texas

GAO project number	DDIR on file	FHWA engineer signature on DDIR	State/ local engineer signature on DDIR	Repair cost estimate	Repair type	Completion date recorded in file for 100 percent funded work	Dates provided show 100% funded work completed in 180 Days	FHWA verification of completion within 180 days	Betterment	Benefit-cost analysis on file for betterment
TX-1	N			N	E	Y	Y	Y	N	
TX-2	N			Y	E	Y	N	N	N	
TX-3	N			Y	P				N	
TX-4	N			Y	P				N	
TX-5	N			Y	P				N	
TX-6	N			Y	E	Y	Y	N	N	
TX-7	N			N	E	Y	Y	N	N	
TX-8	N			Y	E	Y	Y	N	Y	Y
TX-9	N			N	E	Y	Y	N	N	
TX-10	N			Y	P				Y	Y
TX-11	N			Y	P				Y	Y
TX-12	N			Y	P				Y	Y
TX-13	N			N	DK	N		N	N	
TX-14	N			Y	E/P	Y	Y	N	N	
TX-15[a]										
TX-16	N			N	E	Y	N[c]	N	N	
TX-17	N			N	E	Y	N[c]	N	N	

Y Documentation was on file (Y=yes)

Y Documentation was on file but only partially complete

DK Unable to determine the value based on documentation on file (DK=don't know)

N Documentation was not on file (N=no)

Not applicable to that project

E Emergency repair

P Permanent repair

I Incidental permanent repair

Texas

GAO project number	DDIR on file	FHWA engineer signature on DDIR	State/ local engineer signature on DDIR	Repair cost estimate	Repair type	Completion date recorded in file for 100 percent funded work	Dates provided show 100% funded work completed in 180 Days	FHWA verification of completion within 180 days	Betterment	Benefit-cost analysis on file for betterment
TX-1	N			N	E	Y	Y	Y	N	
TX-2	N			Y	E	Y	N	N	N	
TX-3	N			Y	P				N	
TX-4	N			Y	P				N	
TX-5	N			Y	P				N	
TX-6	N			Y	E	Y	Y	N	N	
TX-7	N			N	E	Y	Y	N	N	
TX-8	N			Y	E	Y	Y	N	Y	Y
TX-9	N			N	E	Y	Y	N	N	
TX-10	N			Y	P				Y	Y
TX-11	N			Y	P				Y	Y
TX-12	N			Y	P				Y	Y
TX-13	N			N	DK	N		N	N	
TX-14	N			Y	E/P	Y	Y	N	N	
TX-15[a]										
TX-16	N			N	E	Y	N[c]	N	N	
TX-17	N			N	E	Y	N[c]	N	N	

Y Documentation was on file (Y=yes)

Y Documentation was on file but only partially complete

DK Unable to determine the value based on documentation on file (DK=don't know)

N Documentation was not on file (N=no)

Not applicable to that project

E Emergency repair

P Permanent repair

I Incidental permanent repair

Washington State

GAO project number	DDIR on file	FHWA engineer signature on DDIR	State/ local engineer signature on DDIR	Repair cost estimate	Repair type	Completion date recorded in file for 100 percent funded work	Dates provided show 100% funded work completed in 180 Days	FHWA verification of completion within 180 days	Betterment	Benefit-cost analysis on file for betterment
WA-1	Y	Y	Y	N	P				N	
WA-2	N	DK	DK	N	DK	N		N	N	
WA-3	Y	Y	Y	Y	E/P	N		N	N	
WA-4	Y	Y	Y	Y	E/P	Y	Y	Y	Y	Y
WA-5	Y	Y	Y	N	DK	N		N	Y	N
WA-6	Y	Y	Y	Y	I	Y	N	N	Y	N
WA-7	Y	Y	Y	Y	P				DK[e]	
WA-8	N	DK	DK	N	DK				DK[e]	
WA-9	Y	Y	Y	Y	I	Y	N[d]	N	N	
WA-10	Y	Y	Y	Y	DK	Y	N	N	N	
WA-11	Y	Y	Y	Y	E	N		N	N	
WA-12	Y	Y	Y	N	E/P	N		N	N	
WA-13	Y	Y	Y	N	E/P	N		N	N	
WA-14	Y	Y	Y	N	E/P	Y	Y	N	Y	N
WA-15	Y	Y	Y	N	E/P	N		N	DK[e]	
WA-16	Y	Y	Y	N	E/P	N		N	Y	Y
WA-17	Y	Y	Y	N	E/P	N		N	DK[e]	
WA-18	N	DK	DK	N	DK	N		N	DK[e]	

Y Documentation was on file (Y=yes)

Y Documentation was on file but only partially complete

DK Unable to determine the value based on documentation on file (DK=don't know)

N Documentation was not on file (N=no)

Not applicable to that project

E Emergency repair

P Permanent repair

I Incidental permanent repair

GAO project number	DDIR on file	FHWA engineer signature on DDIR	State/ local engineer signature on DDIR	Repair cost estimate	Repair type	Completion date recorded in file for 100 percent funded work	Dates provided show 100% funded work completed in 180 Days	FHWA verification of completion within 180 days	Betterment	Benefit-cost analysis on file for betterment
WA-19[a]										
WA-20[a]										
WA-21	Y	Y	Y	N	E/I	N		N	N	
WA-22	Y	Y	Y	N	E/P	N		N	N	
WA-23	Y	Y	Y	Y	E/P	N		N	N	
WA-24	Y	Y	Y	Y	E/P	N		N	DK[e]	
WA-25	Y	Y	Y	N	E/P	N		N	Y	N
WA-26	Y	Y	Y	Y	E/I	N		N	N	
WA-27	Y	Y	Y	N	E/P	N		N	DK[e]	
WA-28[a]										
WA-29	Y	Y	Y	Y	E/P	N		N	N	
WA-30	Y	Y	Y	Y	E/P	N		N	Y	Y
WA-31	Y	Y	Y	Y	E/I/P	N		N	Y	Y
WA-32	Y	Y	Y	Y	I	Y	N[b]	N	N	
WA-33	Y	Y	Y	Y	E/I/P	N		N	Y	Y
WA-34	Y	Y	Y	N	E	Y	Y	N	N	
WA-35	Y	Y	Y	Y	E/I/P	Y	N[d]	N	N	
WA-36	Y	Y	Y	Y	E/P	N		N	Y	N

Y Documentation was on file (Y=yes)
Y Documentation was on file but only partially complete
DK Unable to determine the value based on documentation on file (DK=don't know)
N Documentation was not on file (N=no)
Not applicable to that project
E Emergency repair
P Permanent repair
I Incidental permanent repair

Source: GAO analysis.

[a]These projects had been withdrawn by states as they had been determined ineligible by FHWA for emergency relief funds or were reimbursed through a third party insurance settlement.

[b]The information provided included the month and year repairs were completed; without an exact date, we could not confirm completion was within 180 days of the disaster occurrence.

[c]The completion dates provided were beyond the first 180 days after the disaster occurrence and were also for a mix of emergency and permanent repairs, but the state or division remarks field in FHWA's Fiscal Management Information System modifications to the federal-aid project agreements noted completion within 180 days.

[d]The state or division remarks field in FHWA's Fiscal Management Information System modifications to the federal-aid project agreements noted that portions of the repairs were not completed within 180 days of the disaster occurrence and the federal share was modified appropriately.

[e]These project files contained an indication that the project was a betterment—for example, discussion of the project's cost savings to the Emergency Relief Program—although the term betterment was not specifically used, or the scope of work of the project matched the Emergency Relief Manual's description of betterments.

Figure 6. Results of GAO's File Review of Emergency Relief Projects in Three FHWA Division Offices.

End Notes

[1] Federal-aid highways are roads that are eligible to receive federal funding through a series of formula grant programs collectively known as the federal-aid highway program. About 1 million miles of roadway across the country are eligible for federal aid and these roads accounted for about 83 percent of the vehicle miles traveled on the nation's roadways in 2007. The federal government has provided assistance to states in response to natural disasters for many years. The Emergency Relief Fund was established in 1938 (Act of June 8, 1938, 75th Cong., 3d. Sess., ch. 328, § 4, 52 Stat. 633, 634-635).

[2] The Highway Trust Fund is funded on a user-pay principle; it derives revenues primarily from taxes collected on motor fuel and truck-related items and distributes that revenue to the states primarily though a series of formula grant programs collectively known as the federal-aid highway program. The Highway Trust Fund was

created in 1956 (Highway Revenue Act of 1956, Pub. L. No. 84-627, § 209, 70 Stat. 387, 397). Contract authority allows federal agencies to incur obligations in advance of appropriations. A subsequent appropriation is needed to liquidate the obligations. Pursuant to this authority, up to $100 million is authorized to be obligated in any one fiscal year for the program. Any unobligated balance remains available until expended. 23 U.S.C. § 125(c)(1).

[3] GAO, Highway Emergency Relief: Reexamination Needed to Address Fiscal Imbalance and Long-term Sustainability, GAO-07-245 (Washington, D.C.: Feb. 23, 2007).

[4] Pub. L. No. 109-59, § 1112, 119 Stat. 1144, 1171 (2005).

[5] GAO-07-245.

[6] Pub. L. No. 93–288, 88 Stat. 143 (1974), as amended, codified at 42 U.S.C. ch. 68. In September 2011, the Federal Emergency Management Agency issued the National Disaster Recovery Framework which provides a structure for federal, state, tribal, and local agencies to coordinate disaster recovery activities in response to Stafford Act events, which may include emergency relief from FHWA. See Federal Emergency Management Agency, National Disaster Recovery Framework: Strengthening Disaster Recovery for the Nation (Washington, D.C., September 2011).

[7] 23 U.S.C. § 125(d).

[8] 23 U.S.C. § 125(c)(1).

[9] 23 U.S.C. § 125(d).

[10] The criteria for administering emergency relief funds are set out in 23 C.F.R. Part 668. See, also, FHWA, Emergency Relief Manual (Washington, D.C., November 2009). The $700,000 threshold reflects a presumption that lesser repairs should be covered as heavy maintenance or routine repair activities and paid for using other funds. See 23 C.F.R. § 688.105.

[11] The Emergency Relief Manual makes a distinction between current design standards and "betterments," which change the function or character of the facility and are discussed later in this report. The program manual states that repaired facilities may be built to current design standards, which could result in improved or added features that do not change the function or character of the facility. For example, a repaired length of roadway may have wider lanes or shoulders and additional roadside safety hardware that result from following current design standards. According to the manual, these features are not betterments.

[12] 23 U.S.C. § 120(e).

[13] The Emergency Relief Program also funds the repair of roads on federal lands through the Emergency Relief for Federally Owned Roads Program, which is administered through FHWA's Office of Federal Lands Highway. This program is intended to fund unusually large expenses to repair and reconstruct roads and bridges on federal lands that are seriously damaged by a natural disaster or a catastrophic failure. The program may fund 100 percent of the cost of repairs to federal roads.

[14] States with high percentages of federally owned public lands may be reimbursed at a higher federal share percentage, in accordance with a predetermined sliding scale percentage.

[15] 23 C.F.R. Part 668.

[16] FHWA allocates available budget authority to states in a manner that is in some respects similar to the process by which funding is apportioned for purposes of the federal-aid highway formula grant programs. Unlike those programs, however, allocation has no statutory or even regulatory basis. However, as with apportionments, when funds are allocated, cash is not actually disbursed. Instead, states are notified that they have federal funds available for their use. As projects are approved, funding is obligated. Federal funds are only expended when the federal government makes payments to the states for costs as work is completed.

[17] 23 U.S.C. § 106(a)(2),(3).

[18] Only funds for programs authorized under title 23 of the U.S. Code are available for emergency relief work.

[19] 23 U.S.C. § 125(c)(2).

[20] Pub. L. No. 102–240, § 1016, 105 Stat. 1914, 1945 (1991), codified as positive law at 23 U.S.C. § 106(c). See, also, 23 U.S.C. § 106(g) requiring federal oversight.

[21] The National Highway System is approximately 160,000 miles of roadway, and it includes the Interstate Highway System as well as other roads important to the nation's economy, defense, and mobility.

[22] Each state's department of transportation and the FHWA division office in that state have a federal-aid program stewardship and oversight agreement in place which documents the expectations and roles and responsibilities of the state and FHWA in implementing the federal-aid highway program.

[23] GAO-07-245.

[24] GAO-07-245.

[25] GAO-07-245.

[26] U.S. Troop Readiness, Veterans' Care, Katrina Recovery, and Iraq Accountability Appropriations Act, 2007. Pub. L. No. 110-28, title IV, ch. 8, 121 Stat. 112, 160 (2007).

[27] Consolidated Security, Disaster Assistance, and Continuing Appropriations Act, 2009, Pub. L. No. 110-329, div. B, title I, ch. 10, 122 Stat. 3574, 3598 (2008).

[28] Consolidated Appropriations Act, 2008, Pub. L. No. 110-161, div. K, title I, 121 Stat. 1844, 2384 (2008). This supplemental appropriation followed separate legislation passed soon after the bridge collapsed, in which Congress authorized $250 million in emergency relief funds for this project. See Pub. L. No. 110-56, § 1(c), 121 Stat. 558 (2007).

[29] Of the $2.4 billion that FHWA allocated to states from fiscal years 2007 through 2010, about 59 percent ($1.4 billion) was allocated for events that occurred during those years. FHWA allocated the remaining 41 percent ($988 million) for events that occurred from fiscal years 2001 through 2006.

[30] GAO-07-245.

[31] Department of Defense, Emergency Supplemental Appropriation to Address Hurricanes in the Gulf of Mexico, and Pandemic Influenza Act, 2006, Pub. L. No. 109-148, ch. 9, 119 Stat. 2778 (Dec. 30, 2005).

[32] This amount includes $195 million made available through the December 2007 supplemental appropriation.

[33] The National Transportation Safety Board later determined that the probable cause of the collapse was inadequate load capacity of critical bridge components due to an error in the bridge's design. See National Transportation Safety Board, Collapse of I-35W Highway Bridge, Minneapolis, Minnesota, August 1, 2007, Highway Accident Report NTSB/HAR-08/03 (Washington, D.C., Nov. 14, 2008). Emergency Relief Program rules define a catastrophic failure as a sudden failure of a major element or segment of a federal road which is not primarily attributable to gradual and progressive deterioration or lack of proper maintenance.

[34] Pub. L. No. 110-56, § 1(a), (b), 121 Stat. 558 (2007).

[35] In 1996, FHWA amended its Emergency Relief Program regulations to explicitly provide that raising road grades in response to an unprecedented rise in basin water levels was an emergency relief-eligible activity. 23 C.F.R. § 668.109(b)(8).

[36] Pub. L. No. 109-59, § 1937, 119 Stat. 1510 (2005).

[37] App. II provides additional information on Devils Lake emergency relief projects.

[38] 23 C.F.R. § 668.113(a).

[39] The Transportation Equity Act for the 21st Century (TEA-21), Pub. L. No. 105-178, § 1217(a), 112 Stat. 107, 214 (1998).

[40] According to FHWA's Emergency Relief Manual, a pending request is considered an initial estimate of emergency relief needs that is subject to change and is not to result in an actual emergency relief allocation. Instead, pending requests are used to keep FHWA's Office of Program Administration apprised of the most current needs for any known eligible event.

[41] SAFETEA-LU directed DOT, along with WSDOT and the city of Seattle, Washington, to conduct a comprehensive study to determine the specific damage to the Alaskan Way Viaduct from the 2001 earthquake and the amount of assistance from the emergency relief fund for which the viaduct is eligible. Pub. L. No. 109-59, § 1934, 119 Stat. 1482.

[42] FHWA, WSDOT, Seattle City Department of Transportation, *Alaskan Way Viaduct Emergency Relief Eligibility, Report to Congress* (Washington, D.C., June 2007).

[43] The regulatory agenda is a semiannual summary of all current and projected rulemakings, reviews of existing regulations, and recently completed actions of the DOT.

[44] According to DOT's Spring 2011 regulatory agenda, the rulemaking would have considered amending 23 C.F.R. Part 668 to update the annual threshold for an emergency relief event, raise the site threshold and clarify the definition of a site and other definitions, and provide specific time limit restrictions for states when filing a claim for emergency relief eligible work. This rulemaking would also have considered requiring states to develop a plan for obligation needs for emergency relief funding and impose restrictions on the applicability of "quick release" emergency relief allocations.

[45] GAO-07-245. We also recommended that in the event these funds are not needed for other eligible projects, FHWA should identify these funds to Congress either for rescission or to reduce future appropriations.

[46] The $367 million in unused allocations were withdrawn on two separate occasions: on December 1, 2009, FHWA withdrew $105 million, and on November 24, 2010, FHWA withdrew $262 million.

[47] FHWA redistributed this $295 million, as well as other funds made available to the program, through two separate allocations to states. In March 2010, FHWA allocated approximately $339 million to states, which included amounts from the first withdrawal of $105 million from December 2009. In April 2011, FHWA allocated $320 million to states, which included amounts from the second withdrawal of $262 million.

[48] Unobligated funds refer to amounts that have been allocated to states by FHWA, but have yet to be obligated for specific projects. This amount includes all allocations to states as of the end of May 2011.

[49] In total, as of the end of May 2011, $97 million made available to New York for emergency relief in response to the September 11, 2001, terrorist attacks has yet to be used. This includes the unobligated balance of $52 million, as well as an additional $45 million that has been obligated to projects but not expended as of the end of May 2011. This unexpended balance is discussed later in this report.

[50] Unexpended funds refer to program funding that has been obligated to specific projects but has yet to be paid out to states (expended) in reimbursement for completed work.

INDEX

A

B

C

D

E

F

G

H

I

J

L

M

N

O

P

R

S